T0172127

Microsystems and Nanosystems

Series Editors

Roger T. Howe
Stanford, CA, USA

Antonio J. Ricco
Moffett Field, CA, USA

Building on the foundation of the MEMS Reference Shelf and the Springer series on Microsystems**, the new series Microsystems and Nanosystems comprises an increasingly comprehensive library of research, text, and reference materials in this thriving field. The goal of the Microsystems and Nanosystems series is to provide a framework of basic principles, known methodologies, and new applications, all integrated in a coherent and consistent manner. The growing collection of topics published & planned for the series presently includes: Fundamentals • Process Technology • Materials • Packaging • Reliability • Noise MEMS Devices • MEMS Machines • MEMS Gyroscopes • RF MEMS • Piezoelectric MEMS • Acoustic MEMS • Inertial MEMS • Power MEMS • Photonic MEMS • Magnetic MEMS Lab-on-Chip / BioMEMS • Micro/nano Fluidic Technologies • Bio/chemical Analysis • Point-of-Care Diagnostics • Cell-Based Microsystems • Medical MEMS • Micro & Nano Reactors • Molecular Manipulation Special Topics • Microrobotics • Nanophotonics • Self-Assembled Systems • Silicon Carbide Systems • Integrated Nanostructured Materials Structure • A coordinated series of volumes • Cross-referenced to reduce duplication • Functions as a multi-volume major reference work **To see titles in the MEMS Reference Shelf and the Springer series on Microsystems, please click on the tabs "Microsystems series" and "MEMS Reference Shelf series" on the top right of this page.

More information about this series at http://www.springer.com/series/11483

Alissa M. Fitzgerald • Carolyn D. White
Charles C. Chung

MEMS Product Development

From Concept to Commercialization

 Springer

Alissa M. Fitzgerald
A. M. Fitzgerald & Associates, LLC
Burlingame, CA, USA

Carolyn D. White
A. M. Fitzgerald & Associates, LLC
Burlingame, CA, USA

Charles C. Chung
IBM
Armonk, NY, USA

ISSN 2198-0063   ISSN 2198-0071   (electronic)
Microsystems and Nanosystems
ISBN 978-3-030-61711-0   ISBN 978-3-030-61709-7   (eBook)
https://doi.org/10.1007/978-3-030-61709-7

This Springer imprint is published by the registered company Springer Nature Switzerland AG
The registered company address is: Gewerbestrasse 11, 6330 Cham, Switzerland

Foreword

The mass launch of the first real smartphones in the 2005 timeframe marked a major transition for MEMS. No doubt that the discordant MEMS industry was successful in the years leading up to 2005, with crash sensors for airbag systems, the Texas Instruments Digital Light Projector, inkjet printing nozzle arrays, pressure sensors in many applications, and a variety of other devices. But, the 2005 smartphone era marked the first mass consumer introduction of MEMS devices in quantities of 100's of millions. For example, in 2002, there were 2M MEMS microphones shipped. In 2007, there were over 200M MEMS microphones shipped. For the first time ever, in 2008, a MEMS company, Robert Bosch, announced the shipment of its billionth MEMS sensor. In the following years, many companies made that same announcement, and a few years later, several companies were shipping billions of MEMS devices EACH year. Avago, alone, shipped 6B MEMS F-BAR filters in 2019. SiTime, producer of MEMS timing devices, a new product category, has shipped over 1.5 billion devices. In a short period of time, the MEMS industry had transitioned from relatively niche markets into the super high volume consumer markets, with considerably higher stakes.

However, the development and the commercialization of each one of these products was a daunting ordeal, fraught with problems and missteps and blunders and failures. The task of commercializing a MEMS product has been famously well known to be a long, drawn-out affair, taking many, many years, if not a decade or more, before a successful product introduction. While MEMS people will blithely pronounce, "we build CMOS compatible MEMS devices," in fact, a successful MEMS chip can easily take 10 times longer and cost 10 times more to become a commercial product than a comparable CMOS chip.

For example, the first MEMS accelerometers were demonstrated in the late 1970s. They finally came into general production about 15 years later in 1993 for automotive, crash-sensing applications. The first microphone patents were in the early 1990s, coming into production in 2002. The earliest MEMS gyroscopes were demonstrated in 1991, launched into automotive electronic stability systems in 1998. MEMS resonators were demonstrated as long ago as the late 1970s and finally commercialized in 2006, almost 30 years later.

Even though MEMS devices are manufactured by integrated circuit, thin-film processes and technologies, the devices themselves are vastly different from their CMOS circuit counterparts. They are mechanical. They move. They interact with their environment. They mechanically break or stick. In contrast to the two-dimensional geometries of CMOS, MEMS are fundamentally three dimensional. They are more difficult to simulate, which increases the amount of physical trial and error required to design a device with the proper performance. Almost always a new manufacturing process needs to be invented and qualified. They are more difficult to package because of their unique required atmospheres or their interactions with the environment. Their performance over temperature is radically different than CMOS. They are more difficult to test because you also need to apply physical motion to the chip (for an inertial device) or pressure (for a pressure sensor or microphone) instead of just electrical signals for a CMOS chip.

In my career, I have been on all sides of all these issues. I developed our own products and transitioned them to high volume production in our own small MEMS fab at NovaSensor. I have been a fabless MEMS company at SiTime, transitioning our new products to outside foundries, having to establish and maintain relations with these foundry businesses. I have also been on the Board of Directors of a MEMS foundry, IMT, interacting with fabless MEMS companies, who were trying to commercialize their products in our foundry. At every step in the development and commercialization process, the problems are *enormous* and immensely challenging. Even in the best of circumstances, missteps occur and unforeseen complications arise. At NovaSensor, we spent almost a year developing a thin-film, laser-trimmed resistor process for a new pressure sensor. It turned out that, because of packaging requirements, the more cost-effective approach was a thick-film trimming process. The thin-film process was terminated. At SiTime, the product was working perfectly, yields were high. We were weeks away from launching the product. In order to help accommodate our foundry, a committee of some of the best minds in MEMS decided that an incredibly minor process tweak would be just fine. It was a disaster! The mechanical oscillator began to stick. It took 9 months to resolve the problem, and the product launch was delayed by 9 months. At IMT, fully ¾ of the development projects never made it into volume production, mostly for business, not technical reasons.

Finally, "MEMS Product Development," by Alissa Fitzgerald, Carolyn White, and Charles Chung, has come to the rescue! In this incredibly useful book, the mysteries of MEMS development and manufacturing are spelled out in a clear and thorough and incredibly detailed manner. Chapter 1 actually gives us a coherent description and understanding of that question asked above: "Why are MEMS So Challenging?". This book is not another technical description of MEMS devices and fabrication technologies, but, instead, covers real-life, pragmatic business aspects, integration aspects, packaging aspects, testing aspects, how to leverage IP, budgeting, managing costs, selecting a foundry, design for back-end-of-line processes, even the often-overlooked question, "What is the Product?" A question which is ubiquitous in MEMS start-up companies.

What makes "MEMS Product Development" so credible and authoritative is the authors. The team at A.M. Fitzgerald & Associates is indisputably the embodiment of MEMS product development and commercialization. Having completed 100's of MEMS projects with 100's of organizations throughout the world, they are a leader in the pragmatic, real-life world of "getting the job done on time and within budget." I have experienced this exceptional capability myself, having been involved with A.M. Fitzgerald on several projects. Their extensive experience with an enormous breadth and diversity of MEMS devices gives enormous credibility to their advice.

The continuous, ongoing introduction of new MEMS devices and products has not abated. I continue to be astonished by very early demonstrations of completely unforeseen new MEMS devices, usually associated with new start-up companies. So, the potential applications and usefulness of "MEMS Product Development" are huge. I whole-heartedly encourage everyone in the MEMS space who have any glimmer of turning their new innovation into a product, to read "MEMS Product Development." I wish I had access to this book during several of my own companies. I can assure you that it will sit prominently on your desk and will be constantly referenced.

Co-Chair of the HardTech Group   Kurt Petersen
Silicon Valley Band of Angels
San Francisco, CA

Preface

This book is the culmination of more than 17 years of practice and experience in developing novel MEMS devices into commercial products by our company, A.M. Fitzgerald & Associates, LLC. During that time period, we have worked with over 180 companies worldwide and completed more than 400 MEMS development projects, and engineered every type of MEMS device imaginable for a broad range of applications in the consumer, medical, aerospace, and industrial markets. Along the way, we have learned much about both the business and technology of developing MEMS products for commercial sale.

Many excellent textbooks and handbooks already exist on the physics and design of micro- and nanoscale devices and on the processes and materials used in MEMS and semiconductor fabrication. In those technical references, which we cite extensively throughout this book, you may find the deep details on how to design and build MEMS devices.

In this book, we unify that body of technical knowledge with the complex practice of wafer-based product development and the real conditions of manufacturing. Successfully transferring a novel design from its birthplace in the nurturing confines of a laboratory into the machinery of mass production is central to commercialization of every MEMS product.

The book is organized into four sections:

Section I: Context and Overview provides important background material for those who are new to MEMS technology, wafer-based manufacturing or developing novel MEMS products.

Section II: Business Requirements for a Viable Product describes how to analyze the business opportunity and to create a cost model for a MEMS product, in order to validate that there will be a return on the large investment required to reach volume production.

Section III: Technical Requirements for a Viable Product explains the key technical considerations in designing MEMS for volume manufacturing.

Section IV: Technology Transfer and Scaling Up Manufacturing describes how to navigate the leap from a research and development environment to a volume manufacturing facility.

We share our knowledge in order to help innovators, product managers, entrepreneurs, investors, and executives to successfully bring their company's important innovations to market. We hope this book will prove a useful guide for the journey.

Burlingame, CA	Alissa M. Fitzgerald
Burlingame, CA	Carolyn D. White
Armonk, NY	Charles C. Chung

Acknowledgements

We would like to acknowledge our clients and colleagues at A.M. Fitzgerald & Associates, LLC, as well as our many industry colleagues around the world, for countless informative discussions, all of which contributed to the body of knowledge in this book.

We thank Laura J. Stone for providing rigorous editing that helped to make this a more readable text and Korene L. Mangelsen for helping with manuscript organization and submission.

We are grateful to our MEMS colleagues who generously gave their time, as well as their valuable expertise, to provide peer review:

Adi Baram, Brian Bircumshaw, Benjamin Chui, Anthony F. Flannery Jr., Michael A. Helmbrecht, C.T. Kao, Nicole Kerness, Elizabeth A. Logan, Mary Ann Maher, Matan Naftali, Robert M. Panas, Paul Pickering, Magnus Rimskog, James A. Walker, Jason W. Weigold, Paul F. Werbaneth.

Alissa Fitzgerald would like to thank her husband, Alexander Mitelman, for his love, humor and tireless support while so many evenings and weekends were consumed by writing, and our family; her co-authors Carolyn and Chuck for bringing their collective expertise and enthusiasm to this adventure; Julie Ask for her business analyst's critical eye; Tricia Emerson for showing that it is possible to write a book while leading a business; Prof. Tom Kenny, who took a chance on a young graduate student knocking on his office door, thereby launching her career in MEMS; and Kurt Petersen for providing the foreword and for being such an inspiration and mentor to MEMS entrepreneurs.

Carolyn White would like to thank her family, especially David Bianco, for their love and support; Alissa Fitzgerald and Charles Chung for making this book a reality; the educators and mentors that inspired her; her clients for challenging her over the course of her career; her colleagues for making addressing those challenges enjoyable; and the staff at the UC Berkeley Marvell Nanofabrication Laboratory for helping many MEMS development projects succeed.

Charles Chung would like to thank Douglas Henderson for his love, writerly wisdom, and for taking care of our two young kids during a pandemic which made this book possible; Alissa Fitzgerald and Carolyn White who tirelessly wrote and reviewed and edited this book with me; all our friends who reviewed and helped us improve the book; Kurt Petersen for writing the foreword and his generous mentorship to the MEMS community; my colleagues at AMFitzgerald whose insight and congeniality inspire me; and Mark Allen for giving me my start in MEMS many years ago.

Abbreviations

1F	One French (size of a catheter)
2D	Two dimensions or two dimensional
3D	Three dimensions or three dimensional
ADI	Analog Devices, Inc.
AEC	Automotive Electronics Council
ANSI	American National Standards Institute
ASIC	Application specific integrated circuit
ASTM	American Society for Testing and Materials
ASP	Average selling price
BEOL	Back end of line
BOM	Bill of materials
CAD	Computer aided design
CD	Critical dimension
CMOS	Complementary metal oxide semiconductor
COTS	Commercial off-the-shelf
CpK	Process capability key
CTE	Coefficient of thermal expansion
CV curve	Capacitance-voltage curve
DPW	Die per wafer
DRIE	Deep reactive ion etch
EDA	Electronic design automation
FDA	Food and Drug Administration
FMEA	Failure mode and effects analysis
FE	Finite element
FEA	Finite element analysis
GDS	Graphic data system (file format, a.k.a. GDSII)
HAST	Highly accelerated stress test
IDM	Integrated design manufacturer
IMU	Inertial measurement unit
INS	Inertial navigation system
IoT	Internet of things

IP	Intellectual property
ISO	International Organization for Standardization (ISO-9001, ISO-13485)
JEDEC	Joint Electron Device Engineering Council
KGD	Known good die
LIDAR	Light distance and ranging
LPCVD	Low pressure chemical vapor deposition
MEMS	Microelectromechanical systems
MES	Manufacturing execution software
MIL-STD	Military standard
NRE	Non-recurring engineering
OEM	Original equipment manufacturer
OSAT	Offshore assembly and test
PCM	Process control monitor
PCB	Printed circuit board
PECVD	Plasma enhanced chemical vapor deposition
PDK	Process design kit
PFMEA	Process failure mode and effects analysis
POR	Plan of record or process of record
PVD	Physical vapor deposition
ROM	Rough order of magnitude
R&D	Research and development
SEMI	Semiconductor Equipment and Materials International, an industry group
SME	Small and medium-sized enterprises
SoC	System on chip
SIP	System in package
SOI	Silicon on insulator
SOW	Statement of work
SPC	Statistical process control
SSC chip	Sensor signal conditioning chip
TPMS	Tire pressure monitoring system
TRL	Technology readiness level
TSMC	Taiwan Semiconductor Manufacturing Company, Ltd.
TSV	Through silicon via
UV	Ultraviolet
WIP	Work in progress

Contents

About the Authors

Alissa M. Fitzgerald, PhD, founded A.M. Fitzgerald & Associates, LLC, a MEMS product development firm based in the Bay Area, California, in 2003. She has over 25 years of engineering experience in MEMS design and fabrication and now advises clients on the entire cycle of MEMS product development, from business and IP strategy to supply chain and manufacturing operations. Dr. Fitzgerald received her bachelor's and master's degrees from Massachusetts Institute of Technology and her Ph.D. from Stanford University, all in Aeronautics and Astronautics. She is a member of the SEMI-MSIG standards committee and served as a board director on the MEMS Industry Group (MIG) Governing Council from 2008 to 2014. In 2013, she was inducted into the MIG Hall of Fame.

Carolyn D. White, PhD, has a background in mechanics of materials and specializes in the design and fabrication of MEMS devices for a wide range of applications. She has additional experience in technology strategic analysis including of the evaluation of patent portfolios, feasibility studies, and cost/performance analysis. With over 20 years of experience in MEMS, Dr. White is a member of the SEMI MEMS & Sensors Industry Group, previously serving on its Technical Advisory Committee. Dr. White received her B.S. in Mechanical Engineering from Brown University and MS and PhD in Mechanical Engineering from the University of California, Berkeley.

Charles C. Chung, PhD, is the Quantum Industry Consultant for Electronics at IBM. He was at AMFitzgerald for 8 years and has developed MEMS and microsystems products for over 25 years. He is an advisor to the SEMI MEMS and Sensors Industry Group, a member of the University of Pennsylvania's Singh Center for Nanotechnology's Industrial Advisory Board, and a recipient of the Gates Grand Challenges Explorations Grant. Dr. Chung has been involved in the development of over 35 MEMS devices and microsystems products, including DNA sequencing chips, tire pressure monitoring systems, implantable biomedical sensors, quantum sensors, Internet of Things (IoT), micromirrors, microphones, gyroscopes, accelerometers, pressure sensors, and more. His roles include business and product development, technology and patent strategy, device design and fabrication, and supply chain establishment and management for volume production. He received his BA in Physics & Environmental Studies from the University of Pennsylvania, MS in Physics from Duke University, and PhD in Electrical Engineering from the Georgia Institute of Technology.

Part I
Context and Overview

Chapter 1
The Opportunities and Challenges of MEMS Product Development

MEMS devices print our family photos, keep our cars from spinning out of control, deploy airbags in case of a crash, make our smartphones both smart and phones, identify viruses in a patient's sample, and precisely guide oil well drilling heads. Whether as sensors, actuators, microfluidics, optics, or passive microstructures, MEMS are useful in an incredibly wide range of applications: consumer electronics, automotive, medical devices, biotechnology, agriculture, mining, and even aerospace, to name a few.

If you have picked up this book, you probably already have some familiarity with MEMS technology and are harboring some very specific ideas for how to use it in an application that you care deeply about.

You might be a company executive or a product manager, drawn to MEMS technology because you believe it can make your existing products more capable and differentiated in the marketplace. With the latest product trends to use real-time data to personalize experiences or to optimize outcomes, you know that MEMS sensors are essential components for enabling "smart" products and the Internet of Things (IoT) [1].

If you are an innovator or entrepreneur, the infinite possible embodiments of MEMS inspire you to create a novel product that could open up new market applications. You may have formed a start-up company to develop your MEMS design into a component, intended to be integrated into products made by other companies. If so, you have already taken the first step towards commercialization by convincing an investor, or an executive from a future customer company, to finance your product development.

As you are about to embark on your product development journey, you have probably heard a few stories about MEMS commercialization: that it takes many years and many millions of US dollars, and that many companies fail along the way. Those stories are true [2].

© The Author(s), under exclusive license to Springer Nature Switzerland AG 2021
A. M. Fitzgerald et al., *MEMS Product Development*, Microsystems and Nanosystems, https://doi.org/10.1007/978-3-030-61709-7_1

Failure to skillfully navigate the critical step of graduating an emerging MEMS product from its initial research and development environment into the manufacturing environment is a significant source of budget overrun and delay in reaching the market. It can take a heavy toll on an established company, and, for a start-up company, it can be fatal.

We wrote this book to illuminate both business and technical best practices for developing MEMS, especially for transitioning to the volume manufacturing environment. For anyone seeking to commercialize a new MEMS component or MEMS-enabled product—whether innovator, entrepreneur, product manager, executive, or investor—our aim is to brief you on the essential preparations and to empower you with the tools for a successful outcome.

Why Are MEMS So Challenging?

For an industry more than 40 years old, one would think that the pathways to commercialize MEMS would be well paved by now. MEMS have particular qualities, however, that make their development and commercialization more difficult, especially compared to their technological cousins in the semiconductor industry. Some of the difficulties are due to qualities inherent to the MEMS devices themselves, and others to the way that volume wafer manufacturing infrastructure is organized.

Three-Dimensional Features Impede Process Flow Standardization

A MEMS device is a 3D mechanical structure having dimensions in the range of hundreds of nanometers to hundreds of micrometers (microns). These features move, vibrate, or stretch in order to create a specific sensing, actuating, or structural function.

Each micromechanical feature is created on a silicon wafer (or in some cases, a glass wafer) using a sequence of fabrication process steps, organized into a specific order called the process flow. "Surface micromachining" processes, which utilize the same thin film deposition and removal (etch) methods employed in semiconductor device manufacturing, may be used to create some types of mechanical features, typically less than two microns thick. For others, MEMS-industry-specific processes known as "bulk micromachining" are used to add or remove tens or hundreds of microns of material perpendicular to the wafer's surface [3].

In a MEMS device, the desired shape and dimensions of its vertical (Z-axis) mechanical features dictate the fabrication process steps, and, conversely, a certain type of process step is only capable of creating a specific range of vertical features.

The mechanical features within the plane of the wafer (the X-Y plane), however, are not as constrained. Their dimensions are set by photolithographic patterning of the wafer surface.

Designing a MEMS device involves creating a lithography mask layout to outline the X-Y plane features and then selecting a process flow that will define the Z-axis features. If the MEMS device's Z-axis features must change, then so too does the process flow. The MEMS mechanical design and its fabrication process are inseparable.

This is one of several challenges in MEMS: each creative new device design requires a different process flow. The process flow that can create one type of mechanical structure, such as a thin membrane, is not the same as a flow needed to create other structures, such as reticulated, electrostatic comb fingers. The 3D nature of MEMS impedes process flow standardization. This has major impact on manufacturing, discussed further below.

Complex Processes and Microscale Physics Impair Precise Simulation

In a MEMS process flow, the individual fabrication process steps are complex and sensitive to interactions between several variables. Multiple tool input variables such as temperature, chemical concentration or gas mixture ratios, magnetic field strength, and, in some cases, other materials or shapes that are already present on the wafer will affect the outcome of a process step. As of this writing, it is still difficult to simulate with high precision the 3D shape that will emerge from a given process step,[1] let alone an overall new process flow; therefore, predicting a new device's exact behavior from its design is nearly impossible.

In addition, for some types of MEMS structures, microscale multiphysics phenomena may occur during operation. Some examples are squeeze-film damping, anchor energy dissipation, and stiction; each of these phenomena involves complex, nonlinear physics that are difficult to model precisely.

While today's MEMS simulation tools can provide much useful insight and confidence, verifying a new design ultimately requires fabrication of a prototype and careful test and measurement of the design's actual performance under operating conditions. In comparison to other industries, where cycles of learning may be performed entirely within a simulation environment on a computer in a matter of hours or days, MEMS still require a design-fabricate-test iterative cycle. Given the dependencies between design and process flow described earlier, the MEMS cycle is more

---

[1]A priori knowledge of a particular process tool's performance may be input to some MEMS EDA software in order to create a visual rendering of the 3D shape.

like design-(figure-out-the-process)-fabricate-test. When developing a complex new MEMS device, each cycle of learning can take weeks or months—and a lot of money.

MEMS Foundries Must Qualify each New Process Flow

For each novel MEMS device to be manufactured, a MEMS foundry must create and validate (or "qualify") a new process flow. The probability of a foundry being able to someday reuse that process flow to serve another customer's MEMS design is small. Another customer, trying to differentiate its product, will seek to include different features, which then will demand a different process flow. Jean-Christophe Eloy, CEO of Yole Développement in Lyon, France, succinctly lamented the challenges of MEMS development in just four words: "One product, one process."[2]

MEMS foundries have little financial incentive to invest in developing new standardized MEMS process flows to offer to customers.[3] Instead, MEMS foundries charge the customer the cost of developing and qualifying a new custom process flow, in the form of a nonrecurring engineering (NRE) fee. The MEMS company, therefore, must finance the cost and absorb the risk of the foundry's work. Furthermore, any changes to a process flow that the foundry might seek, for example, to improve production yield, must be evaluated for its impact on the MEMS device performance, and vice versa. Developing and qualifying a new process flow for volume production can take years, millions of US dollars, and requires close collaboration between the two parties. When coupled with the limits to precise simulation, rapid new product development is more oxymoron than reality.

Eloy's quip sharply contrasted the plight of the MEMS industry against that of its highly successful technological cousin, the semiconductor industry. While digital ASIC designers, for example, can leverage the semiconductor industry's powerful standardization of transistor process flows and design automation tools to confidently produce high-yielding new chip designs in about 1 year [4], MEMS product companies typically take at least 5 years to launch novel products and may struggle to achieve high yields (Fig. 1.1). As some colleagues sarcastically joke, "MEMS is an acronym for Misses Every Milestone."

---

[2] He later added "and one package".

[3] Some foundries have, however, standardized process flows ex post facto for some common types of MEMS, such as inertial or pressure sensors.

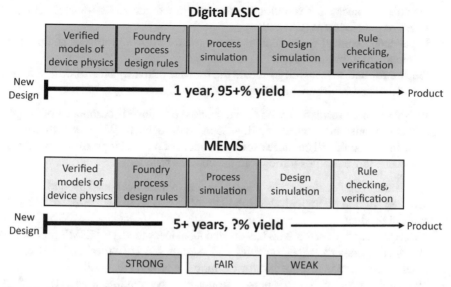

Fig. 1.1 The key elements essential to creating a design automation environment, compared between digital ASIC and MEMS. In the semiconductor industry, process standardization enables high fidelity design automation and creation of new products in less than 1 year; MEMS novel product development typically takes 5 years or more. (Source: AMFitzgerald)

Planning for Successful MEMS Product Development

Despite the challenges surrounding MEMS product development, there are ways to overcome them and to work effectively and cost efficiently. We have learned much while guiding many clients along this journey. Just as you would prepare to have a successful hiking trip through the wilderness by gathering equipment, having a good map, and doing some fitness training, so too must your organization prepare before embarking on developing a new MEMS product. Our book aims to be your guidebook, preparing you to:

- Appreciate the economics of wafer manufacturing and how it influences MEMS manufacturing cost (Chap. 2).
- Understand the stages of MEMS product development and the role of the MEMS foundry in those stages (Chap. 3).
- Accurately define the MEMS product early in development in order to improve future market viability (Chap. 4).
- Create cost models for a potential MEMS product to inform fundraising or financing throughout development (Chaps. 5–6).
- Leverage third-party technology to accelerate product development (Chap. 7).

- Identify the necessary functional roles in your organization during different stages of development (Chap. 8).
- Implement technical best practices for successful MEMS prototype and product development (Chaps. 9–16).
- Prepare for and successfully execute the transfer of a new MEMS design to the manufacturing environment (Chaps. 17–20).

We share this information with an audience we believe is composed of people with backgrounds and interests as diverse as MEMS itself. We expect that some chapters in this book will be more useful to readers of particular profiles than others. We reference those profiles throughout the book as follows:

- "Innovator" or "designer": a person having a technical background who is involved in the technical development of a MEMS device or a MEMS-enabled product
- "Entrepreneur": a person who has created a business to sell a MEMS product
- "Product manager": a person who understands the customer's or end user's requirements and is responsible for defining the product that will be sold
- "Investor" or "executive": a person responsible for funding a MEMS product development and, importantly, for determining whether the development will continue or end.

Summary

This book introduces the vital business and technology components of MEMS product development. The strategies and tactics presented in this book, when practiced diligently, can shorten development timeline, help avoid common pitfalls, and improve the odds of success, especially when resources are limited. We illuminate what it really takes to develop a novel MEMS product so that innovators, designers, entrepreneurs, product managers, investors, and executives may properly prepare themselves to succeed.

References

1. Schadler T, Bernoff J, Ask J (2014) The mobile mind shift. Groundswell Press, Cambridge, MA
2. Petersen K (2014) MEMS entrepreneurial perspectives. In: Solid-State Sensors, Actuators and Microsystems Workshop Hilton Head Island, SC, June 2014
3. Madou M (2011) Fundamentals of microfabrication and nanotechnology, 3rd edn. CRC Press, Boca Raton
4. Nenni D., McLellan P. (2013) Fabless: the transformation of the semiconductor industry. SemiWiki.com, USA

Chapter 2
Economics of Semiconductor Device Manufacturing and Impacts on MEMS Product Development

MEMS was born of the semiconductor industry. But children are similar, not identical, to their parents. Likewise, the MEMS industry inherits much from the semiconductor industry, but has its own unique characteristics, opportunities, and risks. Before embarking on MEMS product development, it is important to understand how the infrastructures for semiconductor manufacturing, in general, and MEMS manufacturing, in particular, impose requirements on MEMS product development and business models.

One reason semiconductor technology has grown to dominate all information processing technologies is its ability to manufacture high volumes at remarkably low per-unit costs. To give a sense of the incredible value that is offered, consider that a microcontroller, a semiconductor chip that has the intelligence to interact with a person and run an appliance, costs about a quarter [1]. A bolt at a neighborhood hardware store costs about the same.

This miraculous value is possible because of relatively low recurring costs in semiconductor manufacturing coupled with huge annual product volumes. Hidden in the product unit cost, however, is the high cost of development and nonrecurring costs such as tooling and capital expenses of infrastructure. Product revenues must be great enough to recover those initial costs over time. For most semiconductor products, such as the microcontroller above, the development costs and capital expenses can be amortized over billions of units to enable a low average selling price (ASP).

Economics of Silicon Wafer Manufacturing

Silicon wafer manufacturing provides terrific value because of high unit volumes. In 2018, global semiconductor unit shipments exceeded 1 trillion units [2]. These remarkable volumes are matched by remarkable investments, which can be in the

A. M. Fitzgerald et al., *MEMS Product Development*, Microsystems and Nanosystems, https://doi.org/10.1007/978-3-030-61709-7_2

tens of billions of dollars for a single manufacturing facility. As incredible as that number is, these costs can be recovered if the product revenues are similarly large or larger.

Parallel Fabrication Enables Economy of Scale

The semiconductor industry is able to make products that offer sophisticated functionality at both high volumes and low prices due to the fact that products are manufactured in parallel (see Fig. 2.1). For most goods in the economy, the unit of work is one, i.e., one unit of the product is worked on at a time at each step of the manufacture. Examples include cars, televisions, and refrigerators. In contrast, in semiconductor manufacturing, many units of the product are worked on simultaneously. The unit of work is the silicon wafer, meaning that each step of the manufacturing process works on a silicon wafer, and on each wafer, there can be hundreds to tens of thousands of products (also called chips or die). All of the die on the wafer are manufactured in parallel, providing a large economy of scale. At the end of the manufacturing process, the wafer is singulated into individual die. Each die becomes a unit of product to be sold. The cost of the die is the cost of the wafer's manufacture divided by the number of die yielded.

To illustrate with some numbers, as of this writing, a typical silicon wafer used in manufacturing is 200 mm in diameter. Suppose a wafer contains 10,000 chips, and it costs approximately $1000 to produce one wafer (a rough order of magnitude cost). Then the cost per product is $0.10 (assuming 100% of the die yield). Many semiconductor wafer manufacturing facilities can manufacture over 100,000 wafers every year.[1] Assuming 10,000 die per wafer, then each of those fabs can produce more than one billion product units each year.

Increasing the number of die on the wafer reduces the product unit cost further. The two ways to do this are to shrink the size of the device or to increase the size of

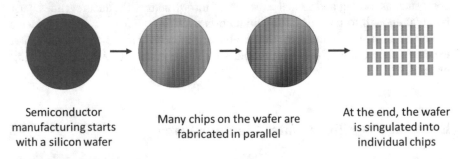

| Semiconductor manufacturing starts with a silicon wafer | Many chips on the wafer are fabricated in parallel | At the end, the wafer is singulated into individual chips |

Fig. 2.1 Semiconductor products are manufactured in parallel on a wafer. The wafer can contain hundreds to tens of thousands of microchips. All of them are manufactured in parallel as the wafer is processed. At the end, the wafer is singulated into individual chips, each one sold as a product

---

[1] The highest volume fabs start over 100,000 wafers *per month*!

the wafer. Over the decades, the semiconductor industry has worked hard to achieve both. Moore's law [3], the observation that transistors would halve in size every 1–2 years, motivated the semiconductor industry for decades to continue shrinking features in order to increase transistor density and therefore enable smaller, cheaper, more powerful chips. In addition, the industry also increased the size of silicon wafers from 50 mm in diameter in the 1970s up to 300 mm in the 2000s. At the time of this writing, 450-mm-diameter wafers are in discussion in the industry.

Large Product Volumes Defray the Huge Capital Cost of Wafer Fabs

The economics of semiconductor fabrication are impressive; however, they require an equally impressive investment in the semiconductor fabrication facility (the "fab") and its equipment. As an example, Taiwan Semiconductor Manufacturing Company (TSMC), the world's largest microchip contract manufacturer, announced in 2017 that it will spend US$20 billion to build a single fab facility for semiconductor production [4].

Fabs are expensive to build for several reasons:

- Due to the extremely tiny, nanometer-scale features being created, the manufacturing environment needs to be ultraclean and precisely controlled.
- The fab must have controlled temperature and humidity and be free of particles, mechanical vibrations, electrical fields, and chemical impurities.
- The input materials to semiconductor fabrication, such as water and chemicals, must be exceptionally pure.
- The fab building itself requires specialized construction techniques, building materials, plumbing, electrical, air-handling, water and air filtration systems, and, due to the hazardous nature of the process chemicals, specialized storage and emergency systems, such as fire suppression.
- Over time, the fab needs continuous maintenance to stay in good working order.

The tools and machinery that populate the fab are also extremely costly. Each tool involved in a single manufacturing operation is an engineering marvel, capable of repeatably producing nanometer structures over hundreds of thousands or millions of wafers per year, operating 24 hours a day, 7 days a week. State-of-the-art production tools range in price from one million to 100 million US dollars per tool. Moreover, each tool requires a squad of technicians and engineers to operate and maintain. With hundreds of tools required to support a semiconductor wafer manufacturing process, it is possible to see why TSMC budgeted US$20 billion for a single fab.

In addition to the facility costs and recurring production costs described above, the cost of developing new semiconductor technology products is significant. Semiconductor fabs invest hundreds of billions of US dollars to develop new processes at ever smaller lithography nodes [5]. Once the process has been developed, the costs to design a chip can exceed $400M for the latest semiconductor processes. For upcoming nodes, design costs in excess of $1B are predicted [6]. After the design is completed, there are then the setup costs to manufacture each product design.

These costs are incredibly high, but the semiconductor industry can make these investments because product revenues will recoup the costs. In 2019, worldwide revenue from the sale of semiconductor chips exceeded US$418 billion [7].

Captive Fabs, Foundries, and the Fabless Model

Companies that have a high volume of products and can fully utilize a fab's capacity will own and operate a fab solely for their own products. These are called "captive fabs." Companies with captive fabs are called "integrated device manufacturers" (IDM), and they include Intel, Texas Instruments, Micron, STMicroelectronics, Bosch, and Samsung.

But many companies cannot fully utilize a fab, or some lack the capability to manage a fab, or some prefer not to invest in ownership of a fab. Instead, these companies contract with a "foundry," which is an entity that owns and operates fabs solely to fabricate other companies' chip designs. Companies that design semiconductor or MEMS products, but do not own their own fab, operate under what is called the "fabless" model. Companies of all types utilize the fabless model, including Apple, Google, Nvidia, AMD, and many more.

To help their customers' designs achieve the best yield and performance, the foundry usually provides a set of process design rules, process data important for design, and other guidance for the customer's design process. The customer sends only its design data to the foundry, and then the foundry fabricates and delivers the completed wafers to its customers.

The captive and fabless models are not mutually exclusive. Companies with captive fabs may also process some of their products at a foundry, especially if the foundry can provide processes or throughput not available in their captive fab. Conversely, companies with captive fabs may not have the product volume to fully utilize their fab, so they may open their manufacturing capacity to other companies.

TSMC pioneered the pure manufacturing foundry business model in the mid-1980s, and the semiconductor industry embraced it [8]. A key innovation of the foundry model was to enable each party to create and maintain its own distinct intellectual property; the foundry invests in its own processes, and the customer develops its own designs. This segmentation enabled smaller, design-only companies to proliferate, since they need not undertake the huge capital investment of owning and operating a fab, but retain access to state-of-the-art semiconductor processes. As of this writing, the foundry model has become so popular that now more than 200 foundries operate around the world. And nearly 40 years after pioneering this model, TSMC still remains the largest foundry in the world [9, 10].

How MEMS Production Differs from Semiconductor Production

The MEMS manufacturing infrastructure and supply chain inherits some structural similarities to semiconductor manufacturing, but also has its own distinguishing characteristics. Understanding these differences enables a product development team to plan for the unique characteristics of MEMS production.

MEMS Devices Are Physically Different from Electronic Semiconductor Devices

There are several MEMS devices in every smartphone and dozens in every automobile. These devices enable your smartphone to listen to your voice, improve camera images, and provide directions to your destination. In your car, they improve fuel efficiency, sense collisions, and assist on slippery roads.

To achieve these capabilities, MEMS work with materials and dimensions that are uncommon in semiconductor electronics. Examples of MEMS-specific materials include noble metals, piezoelectric materials, magnetic materials, structural metals, structural polymers, and many more. These materials provide unique capabilities to MEMS devices and require a different set of tools than would be found in a semiconductor fab.

Another unique "material" is the empty space in MEMS devices. Most semiconductor devices are monolithic; they have no empty spaces. MEMS devices, in contrast, move, and therefore require free space around their structures. Formation and treatment of these open spaces require special consideration in their design and fabrication. For successful MEMS fabrication and operation, those spaces must remain free of any unintended materials, such as dust, debris, or water.

MEMS also use a wider range of geometries to achieve their capabilities than semiconductor electronics products (see Fig. 2.2). The in-plane shapes of electronic devices are largely composed of rectangles. In contrast, the in-plane shapes of MEMS devices are unrestricted, and structures may have complex shapes, such as curves, protrusions, and reticulations.

Also, electronic devices are mostly planar, typically extending less than a few micrometers in the vertical direction (recently, there are exceptions to this, such as 3D ICs). In contrast, MEMS structures typically have prominent features in all three dimensions. Some MEMS devices have features that extend into the vertical dimension by hundreds of micrometers.

To shape these unique materials into unique geometries, MEMS fabs utilize tools not found in a semiconductor fab. These include tools which perform processes such as front-to-backside lithographic alignment, deep reactive ion etching, vapor hydrofluoric acid etching, and wafer bonding. Chap. 11 provides more detail on these processes.

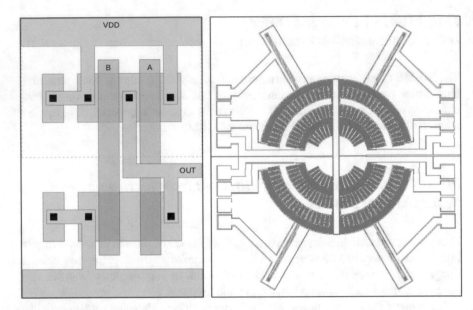

Fig. 2.2 Left: Layout for a typical semiconductor electronic device (CMOS NAND). Right: Layout for a MEMS device (rotary optical positioner). The electronic device uses only rectangular geometries. The MEMS device uses more complex geometries, including curves and arbitrary angles. Left image is public domain from Wikipedia Commons [11]

In addition, MEMS devices have different lithography needs from other semiconductor devices. For example, many MEMS can be made using lithography tools at the 0.35-um or 1.0-um nodes, which is an older, larger, and cheaper technology compared to semiconductor lithography tools manufacturing at 14-nm nodes or smaller. In addition, due to its three-dimensional nature, MEMS lithography often needs to be performed over nonplanar surface topology. Finally, MEMS devices may need to be aligned to features on the opposite side of a wafer, which is atypical in semiconductor processing.

MEMS Lack Standard Processes

While the semiconductor foundries invest heavily in standardized processes to offer to their customers, the MEMS foundries have had little incentive to do so. Semiconductor electronics products are based on combinations of a few element devices, the most common being: transistors, resistors, capacitors, and diodes. Different electronic chip designs differ by the particular combinations and sizes of these elements. As a result, a standard foundry process that can produce these elements can produce many different types of electronic products: microprocessors, graphics processors, artifical intelligence chips, amplifiers, filters, power regulators, and many more. A standard process justifies the massive investment because it can be shared among so many products. A fabless electronics company needs only to supply a device design to the foundry, and need not develop a process to manufacture its products.

MEMS products, on the other hand, are not composed of combinations of elemental devices. As described in Chap. 1, each MEMS product is its own device with its own process. As a result, the design and process support services and datasets available at semiconductor foundries to support new product development are available only for a few, mature MEMS devices, such as accelerometers. This remains true even today, some 30 years after MEMS foundries first emerged.

With each MEMS product requiring a new process, a MEMS company must budget and plan to develop both the device design and its process. The process development cannot usually be shared with other products or other companies and the process usually takes years to develop and then scale for volume manufacture.

MEMS Products Have Lower Wafer Volumes per Process

For MEMS products, the volume of wafers *per process* is generally lower than semiconductor products such as microprocessor or memory chips. There are several reasons for this. One reason, as discussed in the previous section, is that many semiconductor products can share the same standard process. In contrast, MEMS processes tend to be unique to each MEMS product.

Moreover, many MEMS devices must be engineered for their specific end use application. This specialization and customization fragments wafer manufacturing volume, which can result in a MEMS product company dividing its wafer volume among multiple products, such as for a family of pressure sensor products.

Finally, MEMS chips are typically smaller than many semiconductor electronics chips. MEMS chips tend to be smaller than 2×2 mm; microphones and pressure sensors may be smaller than 1×1 mm. A 200-mm wafer has space for approximately 30,000 chips that are 1×1 mm in area. Semiconductor electronics products, such as a memory chip or a microprocessor, are often bigger, such as 5×5 mm or larger. A 200-mm wafer would therefore have space for only 1200 5×5 mm chips. To produce the same annual volume of die, a semiconductor product will need to fabricate 25 times more wafers than a MEMS product, assuming equivalent wafer yields.

The result is that, generally, fewer wafers are manufactured for a MEMS process, so the amortization of the foundry's process development costs that is enjoyed by semiconductor wafers may not be possible for MEMS wafers. Consequently, MEMS process development costs may be a substantial fraction of the total cost of a MEMS wafer and lead to a much higher per wafer cost. The MEMS foundry process development cost, the expected wafer pricing, as well as the total manufacturing costs and unit volumes, should be carefully evaluated to determine that a MEMS product's per-unit price will insure a positive return on investment.

Summary

When navigating a development path for a new MEMS product, understanding the general economic drivers of the semiconductor industry offers important guidance, and understanding when and how those patterns are altered for MEMS is crucial.

Economy of scale is only realized at high annual volumes of wafer production. One cannot assume that typical semiconductor industry economics will be applicable to a MEMS product because they can differ in terms of processes, wafer volume, and many other aspects.

When planning for MEMS product development and for a company business model based on a MEMS product, it is critical to keep these key industry differences front of mind. Teams who have deep experience in the semiconductor industry but who are new to MEMS development often have difficulty in identifying their assumptions about semiconductor manufacturing and are especially vulnerable to misunderstanding MEMS manufacturing economics.

References

1. Microchip Technology. Part no. ATTINY4-TS8R (2020) Mouser. https://www.mouser.com/. Accessed 31 May 2020
2. (2019) Semiconductor Unit Shipments Exceeded 1 Trillion Devices in 2018. IC Insights. https://www.icinsights.com/news/bulletins/Semiconductor-Unit-Shipments-Exceeded-1-Trillion-Devices-In-2018/ Accessed 31 May 2020
3. Mack C (2011) Fifty years of Moore's law. IEEE Trans Semicond Manuf 24(2):202–207
4. Yu JM (2017) TSMC says latest chip plant will cost around $20 bln. Reuters. https://www.reuters.com/article/tsmc-investment/tsmc-says-lat-est-chip-plant-will-cost-around-20-bln-idUSL3N1O737Z. Accessed 16 May 2020
5. LaPedus M (2019) 5nm Vs. 3nm. Semiconductor Engineering. https://semiengineering.com/5nm-vs-3nm/. Accessed 30 July 2020
6. LaPedus M (2018) Big trouble at 3nm. Semiconductor engineering. https://semiengineering.com/big-trouble-at-3nm/. Accessed 30 July 2020
7. (2019) Gartner says worldwide semiconductor revenue declined 11.9% in 2019. https://www.gartner.com/en/newsroom/press-releases/2020-01-14-gartner-says-worldwide-semiconductor-revenue-declined-11-point-9-percent-in-2019. Accessed 25 August 2020
8. (2013) A fab success. The economist. 27 July 2013
9. Manners D (2019) Q3 foundry revenue to rise 13%. Electronics Weekly. https://www.electronicsweekly.com/news/business/q3-foundry-revenue-rise-13-2019-09/. Accessed 28 May 2020
10. Nenni D, McLellan P (2013) Fabless: the transformation of the semiconductor industry. SemiWiki.com, USA
11. (2006) Wikipedia. https://en.wikipedia.org/wiki/NAND_gate#/media/File:CMOS_NAND_Layout.svg. Accessed 26 August 2020

Chapter 3
Stages of MEMS Product Development

In this chapter, we assume that your company has already invented an interesting new MEMS or microsystems technology and now seeks to turn that technology into a commercial product. We define five stages of development that your MEMS will pass through on its way to volume production. Different companies and foundries may use varying terminology to name and group these stages, but the overall concepts presented will apply and are designed to help you to smoothly transition from one stage of development to the next while avoiding common pitfalls. This chapter provides a big-picture overview and context for the information provided in the following chapters.

The Five Stages to Commercialize a MEMS Product

We define five stages of MEMS product development, referenced throughout the book (Fig. 3.1):

- Proof-of-concept prototype.
- Advanced prototypes.
- Foundry feasibility.
- Foundry pilot production.
- Foundry production.

We define the stages according to the challenges that MEMS-based products must address in each. For this reason, the stages are defined in terms of the maturity of the MEMS component only, but of course, would include any accompanying component or system development needed to make the product function, such as ASIC, package, or software.

In the proof-of-concept prototype stage, you must convince an investor or executive that your novel product idea is worthy of further investment, often while having only a small budget. This work could take place in an academic environment, an

A. M. Fitzgerald et al., *MEMS Product Development*, Microsystems and Nanosystems, https://doi.org/10.1007/978-3-030-61709-7_3

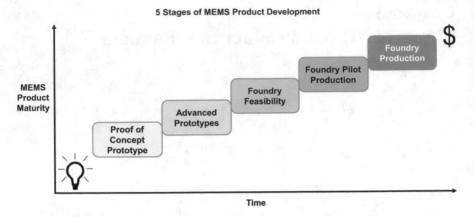

Fig. 3.1 The five stages of development to commercialize a MEMS product

early stage start-up, or in a large company that has its own R&D department. In the advanced prototypes stage, you must establish that the device exhibits the desired performance and has a process that can be fabricated at the required volumes, at the right cost. The foundry feasibility stage is where your chosen foundry confirms its ability to make your MEMS device. Once confirmed, the foundry moves into the pilot production stage where the process flow goes through characterization and qualification[1] to achieve the best yield and device performance, and then the process flow is repeated without changes to gather statistics. Once this qualified process flow demonstrates that the devices consistently meet yield and performance specifications, you are ready to move into the foundry production stage and volume manufacturing.

Proof-of-Concept Prototype

In the proof-of-concept prototype stage, the first physical demonstration of the function of a new technology, and a potential product, is done. During this stage, you will be figuring out how exactly your technology works, and how it may eventually be optimized to meet a potential customer's needs. Prototypes in this stage are usually crude, much larger than a potential product could ever be, fragile, and created by inefficient methods and manual labor. Usually they are produced in quantities of only 1–10 units at a time, or of one or two wafers at a time. They are vehicles for understanding the technology and not yet actual products.

Thinking you are ready to go to foundry production once the first few working prototypes show promise is a common pitfall. While results from these devices may

---

[1] In this book, we have assumed that process qualification occurs at the start of the pilot production stage. At some foundries, this is done at the end of the feasibility stage and the pilot production stage consists only of fabrications runs based on the final process flow.

be sufficient to publish a research paper or for a pitch meeting with an angel investor, it is not enough upon which to immediately build product. We have worked with many clients who licensed technology developed in an academic setting and who assumed that those early prototypes could be used to immediately begin product or system development. It came as quite a shock to their business plans when we informed them that the proof-of-concept process flow would not be manufacturable at their expected product volumes or that we needed to do a complete process redesign in order to make the device compatible with typical foundry manufacturing capabilities.

In this development stage, what is most needed is an ability to make proof-of-concept prototypes quickly, cheaply, with easily accessed resources and perhaps even a few short cuts. Frequent, short cycles of learning are essential to efficiently explore the technology's potential as a product. The guiding principle for the proof-of-concept prototype stage is to focus on what level of device performance you need to develop and demonstrate in the near term, while still keeping in mind the requirements and constraints of later stages of development.

You still have latitude for experimentation in this stage. For example, you might build prototype chips large enough to manually probe and package (see Chap. 11 on planning for future die shrink). You may use a thick, manual photoresist coat process that can accommodate your wafer's topography. Or you may manually cleave your wafer into quarters in order to use an older generation tool that will not fit a full-sized wafer. But no matter what, always have a path in mind for how that same process step could later be done in a production environment.

Once you have a working prototype and a fuller understanding of your technology, then it is time to shift your attention to getting the resources you will need to further develop your product in the advanced prototype stage.

Advanced Prototypes

The advanced prototypes stage is the primary focus of this book. It is where much critical product development work needs to occur as you further optimize device performance and develop its design and process for compatibility with foundry production. In this stage, your company's product development mindset must shift from "could it succeed?" to "how do we make sure it will definitely succeed?" There will be multiple design iterations in this stage, which is why we call it advanced prototypes - plural.

In the advanced prototypes stage, design for future manufacturing at a foundry becomes critical. You have the greatest flexibility for design changes throughout this stage, so use it wisely, because that flexibility will disappear in later stages. Your prototyping efforts must now focus on how to make the MEMS device and the surrounding system in quantities of hundreds to thousands of prototypes versus tens. Expect to process multiple, small batches of wafers (e.g., 5-10 wafers for each process flow run) in order to systematically iterate and validate device design, fabrication process flow, generate demo or sample units that meet the product speci-fications, and provide sample units for building prototypes of the entire product or system.

In addition, during the advanced prototypes stage, you will:

- Perform cost analyses of different process flows and materials to make sure engi-neering choices do not invalidate key business model assumptions (see Chaps. 4–6).
- Research available foundry platform process flows or licensing opportunities for possible development shortcuts (see Chap. 7).
- Examine the compatibility of your materials and process flow with what is avail-able at MEMS foundries (see Chap. 11).
- Determine how the MEMS device will be packaged and develop the surrounding system and test plan (see Chaps. 9, 12–14).
- Examine if regulatory approval and qualification testing is necessary for the end use application and how this will affect the product launch timeline (see Chap. 14).

You must also have the business case to justify your on-going product develop-ment and the funding in hand before going too far into the advanced prototypes stage. This is the stage at which a significant investment is required and money will be spent at a rate of US $100,000s–US $1,000,000s/year. Section II of this book provides guidance on evaluating business viability. Section III of this book devotes multiple chapters to addressing technical aspects of development.

The combination of the technical and business cases, and also the resources available for your MEMS product, determines how long you might stay in the advanced prototypes stage. Without this stage's preparation, you will find it difficult to navigate the remaining product development stages. With some planning and smart decisions, the groundwork in this stage can decrease the time spent in the remaining stages, where the cost and time to recover from failures is much higher.

A number of companies and organizations provide services well suited for this stage (see Table 3.1; more information regarding development foundries and pro-duction foundries can also be found in Chap. 17). Smaller development organiza-tions may be better suited to do the small wafer batch processing that the advanced prototypes stage requires because they tend to be nimble and less expensive. However, there will likely still be a need to transfer the final design to a volume foundry once development is complete. Larger development organizations may be able to provide seamless continuity between development and production stages, if their fab is large enough to your handle production wafer quantities. These organiza-tions must be motivated to work with your company while you still need small wafer batches of prototypes, which often means that they will first want to be sure that your product has the potential for larger wafer batch orders. Motivation might be based on your prior success in developing a MEMS product, their assessment of your device, or known market pull and your ability to capture a significant market share.

In Table 3.1, we also identify which types of development organizations are typi-cally best suited for serving different product company types. Choosing where, and with whom, to develop your advanced prototypes depends on the type of MEMS device, the end use application, the market, the product or system requirements,

Table 3.1 Types of organizations that can perform advanced MEMS prototype development

Organization type	Organization characteristics	Typically best suited for…
Government-backed R&D facility	• Often large organizations with access to both established and experimental tools • May require joint IP ownership for any developed technology • Often supported by local government to promote economic growth • Production may require transfer to a larger fab	• Established SME product companies • Large product companies • Product companies lacking internal MEMS expertise
Development service providers	• Experienced in the challenges of MEMS development • Can provide both design and wafer processing services • Can assist with transition to volume production • Requires transfer to foundry for production	• Start-ups • Established SME product companies • Large product companies • Product companies lacking internal MEMS expertise
Development foundries	• Able to process smaller volumes, wafer splits, etc. during advanced prototyping • Usually processes smaller wafer diameters (150 mm or less) and has an older tool set • Focused on wafer processing, including process development, limited or no design services • May require transfer to larger fab for production	• Start-ups • Established SME product companies • Large product companies • Companies having good expertise in MEMS
Production foundries	• No design services available • Usually process on larger wafer diameters (150 mm or more) • Usually requires combination of significant wafer volume forecast and strategic business alignment • Can support entire production volume	• Very well-funded start-ups • Established SME product companies • Large product companies • Companies with high expertise in MEMS

your company's structure, expertise and funding, as well as the organization's willingness to work with your company.

For discussion purposes in this book, we have assumed that your company will not be developing its advanced prototypes in a production foundry; however, the concepts we present still apply no matter at which organization or facility the work is actually done. If you are new to the MEMS development ecosystem, this book will help you to assess all of these factors and enable you to make an educated decision on how to pick partners for the advanced prototypes stage.

As you reach the end of the advanced prototypes stage, Chap. 16 will guide how to assemble the documentation you will need to take the first step towards production at a volume MEMS foundry. Chap. 17 can help determine if your product and your company are ready to make that technical and business leap. Transitioning to the production fab environment with a well-documented design and process already in-hand will greatly increase your chances of efficient and successful manufacturing ramp up.

Foundry Feasibility

Due to the significant investment both you and the foundry will be undertaking to develop the production process, it is important to select the right foundry for your near-term and long-term goals[2]. Information on this crucial selection process can be found in Chap. 18.

Once you have completed development of your advanced prototypes and are ready to transition to a production facility, you will be taking a step that only a small percentage of proof-of-concept prototypes ever achieve. No matter how many prototypes you have previously built, however, the minute you move to a MEMS production fab, you start at a new square one for your product: foundry feasibility. Here the term "feasibility" does not reflect whether your device can be made or not—your advanced prototypes have already shown that it can.

Instead, you are determining the feasibility of making your device at a particular foundry, on its particular set of tools. As a result, there should be no expectation that the first or even second feasibility run at a foundry will produce working devices. This applies even if you are transferring a design that has already been in production at another foundry (that is, has already moved through the feasibility and pilot production stages at the other foundry). MEMS processing is extremely dependent on the specific tools in a fab. Differences can even be seen between two tools that are the exact same make and model that were installed in the same foundry at the same time [1]. Process variation is a fact of life in MEMS processing. A process must be checked and re-qualified any time even the smallest change is made.

The first time a foundry runs a new process (your transferred advanced prototype process), a foundry project manager supervises a small batch of wafers through all the process steps. It is the manager's job to adapt your process to the foundry's toolset, evaluate each process step, and troubleshoot process issues when (not if) they arise. The foundry may elect to run a series of short loops to prove a process module or two before committing a batch of wafers to a complete process flow (see Chap. 11 for more information on process modules vs. process flow). The foundry will base these decisions on the information provided by you about the advanced prototype design and process (see Chaps. 16 and 19) and its own expertise. Well-organized, detailed design and process flow documentation on your advanced prototypes will save you time and money because it will reduce risk and assist the foundry in process transfer.

If you might have the opportunity to use a MEMS foundry's platform process[3] to fabricate your MEMS device design, you will be able to skip over the proof-of-concept and advanced prototypes stages, because the foundry has essentially com-

---

[2] In this book, we have generally assumed that the wafer production for a MEMS product company will take place at a MEMS foundry. However, the concepts and advice provided in this book would also be valid for products made within a company's captive MEMS fab (see Chap. 2 for more information on captive fabs vs. foundries). Many large integrated companies run their captive fab as a separate business unit and would regard you as an internal customer.

[3] A MEMS foundry platform process is a foundry-defined process flow for a given type of MEMS device, usually a pressure sensor or motion sensor, with accompanying design rules that must be rigidly followed.

Foundry Feasibility	**Foundry Pilot Production**	**Foundry Production**
• Transfer of your MEMS process to the foundry • Determines the feasibility of making your unique device at the foundry not the feasibility of the device itself • Often includes short loops and other risk mitigation steps	• Gated by finalized MEMS design and process flow • Proves process can be reliably run without constant oversight • Collects data to determine the allowable variation in production that meets device performance and yield criteria	• Gated by the customer's criteria and mutually agreed upon yield • Regular production of product quality wafers that meet agreed upon acceptance criteria • Ongoing relationship needs support for quality monitoring and process engineering for continuous improvements

Fig. 3.2 High-level summary of typical foundry product development stages

pleted those stages for you. You will, however, still need to go through each of the foundry stages, but can do so at a much faster pace. When using a foundry platform process, the feasibility stage serves to confirm that your design can be made on the platform process with the expected performance.

We briefly touch on foundry platforms in Chap. 11, but for the most part, we assume throughout this book that you are developing a custom MEMS process flow to fabricate your product. At the time of this writing, using foundry platform processes is not yet a common path for novel MEMS product development because of the limited number available and their limited versatility.

Once the foundry has completed feasibility wafer runs that consistently meet the customer defined performance criteria and there are no further modifications to the device design, mask layout, or process flow needed (also referred to as a "frozen" design or process flow), the foundry will release the process to the next development stage, pilot production. The exact criteria for leaving the foundry feasibility stage may depend on additional factors including expected yield, customer cost requirements, and so on.

Each foundry has its own terminology and groupings for its product development stages (the names engineering, prototype, qualification, and pre-production are also commonly used) but the goals and action items for the overall development flow is the same. Figure 3.2 provides a high-level summary of the typical foundry stages that we have defined in this book.

Foundry Pilot Production Stage

The purpose of the pilot production stage is to confirm that the frozen design and process flow can be run in a production mode, that is, by operators and without constant oversight of an engineer. It is also used to further tune the process to produce higher yields and to collect data to determine the allowable variation in each process step.

In order to practice and perfect the wafer fabrication process, some foundries may choose to run process variants to explore the limit of their tools' capabilities to meet the agreed-upon product pass/fail criteria. This work is not focused on design improvement but on understanding and documenting which process tolerances are essential to producing high-quality product and how to maintain the tool's performance at that optimal setting. Some foundries refer to this as "process windowing" or running "corner lots." With each wafer batch fabricated, the foundry gathers more data to monitor the process and to gradually improve the yield of the wafer [2].

A larger quantity of wafers must be processed in this stage in order to support these efforts as well as to assess what is needed to meet your production volume requirements. This may include moving process modules from a vendor to an in-house tool or investing in additional equipment. In this stage, the time it takes to complete each run should also be improved. Shorter run cycle time means you can identify and fix process issues faster and achieve an increase in production volumes within a certain time period.

The wafers fabricated in this stage will likely have multiple uses beyond helping the foundry to perfect the process. They should be a high enough quality to provide product samples to your customers, or to be supplied to other vendors in your production supply chain to establish their own downstream processes and protocols (see Chap. 13). If your development of the product packaging, electronics, and the system performance testing all match pace with the fabrication of the pilot production MEMS wafers, then first product sales could possibly begin during this stage and thereby provide revenue to fund your ongoing product development efforts.

Timing of this stage is important. The foundry pilot production stage must occur just prior to the start of volume production. Any gap in time between the completion of this stage and the start of volume production can lead to the foundry's carefully calibrated process windows drifting with time. Operator expertise can also become dulled and staff reallocated. If long delays were to occur, such as those due to changing market conditions or lack of funding, be prepared to have to repeat this stage. If more than a year has elapsed since your last wafer run, you may need an additional feasibility stage. Be realistic in your business planning and financing in order to avoid wasteful gaps in the ramp-up to manufacturing (see Chap. 17).

Foundry Production

You will reach the final stage, foundry production, when the foundry is confident it can reliably fabricate wafers that will meet a set of previously defined product acceptance criteria. The foundry is now ready and able to fabricate thousands of high-quality product wafers at a continuous pace.

A regular production schedule will be established in order to meet your company's annual supply and inventory demands. The foundry will sell a certain volume of wafers, for a negotiated per wafer cost, to your company. Your company's interaction with the foundry will now involve regular purchasing transactions, monitoring quality control data, and engineering support for continued improvements, process issues, and process tool changes.

Foundry production is the end stage of MEMS product development. Reaching it is a major organizational achievement and cause for celebration. However, once achieving production, remaining at this stage is not guaranteed, and furthermore, some events could cause you to regress from this stage.

For best results, wafer production needs to occur continuously. Any large gaps in time between orders could result in process drift; complex processes having tight process tolerances go stale the fastest. A gap of only a few months with no wafers being processed could send you back to the pilot production stage to re-qualify the entire process. Planning and communication with the foundry are essential to manage this risk should your company experience a drop in demand for product.

Other events that could kick you back to the pilot production stage would be any changes in chemicals, materials or tools used. Suppliers change their product offerings all the time, tools become obsolete, and the foundry may be forced to adapt to these changing circumstances. In addition, at some point the foundry may want to upgrade its tool set to accommodate a larger wafer size, in order to improve its business capacity and profitability. This is a massive undertaking that would involve tens of millions of US dollars and take several years to execute. It would be carefully planned and coordinated with customers as certain tools are removed and new ones brought in. Moving a process from one wafer size to the next (such as from 150 to 200 mm) involves tools having newer process technology. This means that your previous production-ready process will be pushed back to the pilot production stage and have to go through the process windowing exercise again on the new tool set.

So even though the foundry production stage is the end state of development, it is also a state which needs continual effort and management to maintain over time.

Funding and Schedule for the Five Stages

Over time, the funding required to develop MEMS products will add up to a large sum. If you approach a MEMS product development with calibrated expectations, you can prepare accordingly, establish the right technical milestones, secure sufficient budget, and most importantly, also manage the expectations of investors and executives along the way.

Typical Budget and Timeline for Stages of Development

As with most things in MEMS, the cost of development depends largely on the specific device design, process, and additional components in the product. Chaps. 4–6 provide detailed guidance to help you determine the budget needed to develop your MEMS. Historically, MEMS product development budgets have ranged from a few millions to tens of millions of US dollars, with a few well over US$ 100 million.

For the purpose of calibrating expectations, Table 3.2 provides a minimum budget for each development stage of a MEMS device having medium complexity, based on our business experience. By medium process complexity, we mean a process having 4–7 mask layers, wafer-level electrical testing, and standard saw dicing. Table 3.2 also includes budget for your engineering and product development teams, a purchased ASIC component, off-the-shelf packages, and vendor services for back-end processing, assembly, and test. It does not include other concurrent business operating expenses such as overhead, accounting, sales and business development, or capital expenses (e.g., equipment, software).

We present this as a minimum development budget, because it is certainly possible to spend much more, and higher complexity products (more complex MEMS and/or electronics, or a higher degree of integration) will definitely consume a lot more budget.

Figure 3.3 shows these budgets along with a typical timeline for each stage, based on our business experience serving multiple clients. Again, this figure represents a minimum for a MEMS device of medium complexity. Please see Chap. 6 for additional information.

Table 3.2 Minimum engineering budget per development stage for a new MEMS product of medium complexity (please see Chap. 6 for a more detailed analysis)

Development stage	Minimum engineering development budget
Proof-of-concept prototype	US$ 500 K
Advanced prototypes	US$ 1.5 M
Foundry feasibility	US$ 1 M
Foundry pilot production	US$ 1 M
Total	**$4 M**

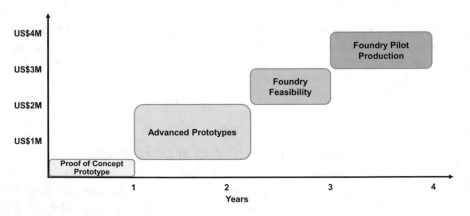

Fig. 3.3 Minimum development budget and timeline overview for a MEMS device of medium complexity. Includes engineering work only; no overhead business expenses

Our intent is not to scare you, but to help prepare you. MEMS product development is a long and expensive undertaking; knowing what you are about to get into is essential to improving your odds of success.

Secure Continuous Funding for Efficient Product Development

As shown in Fig. 3.3, MEMS product development is an undertaking that requires millions of dollars of funding sustained over several years. It is a lot to ask of any investor or executive. Cautious funders will attempt to manage their exposure to risk by making each new round of funding dependent on a set of milestone achievements and go/no-go criteria, where the milestones are usually points within the development stages that we have defined in this chapter. Problems will arise if your company is able to successfully pass the development milestones and yet onward funding for the next stage is not immediately available.

Discontinuous funding is a common issue we have frequently seen plague our customers' MEMS product development. Whether funding comes from internal company R&D funds, venture capitalists, or government grants, you can be caught unfunded in between development milestones due to a lack of planning or events beyond your control. A common cause is simply insufficient communication with the company CFO or investors regarding when you need the funding to arrive and the consequences and additional costs that would be incurred by any delays. Circumstances beyond your control, such as economic downturns or timing of grant awards, can also cause major impact.

Throughout MEMS development, any downtime that arises between development stages will lead to serious inefficiencies that, in the end, will add significant cost to the overall product development. It takes a lot of effort and money to build momentum, due to the number of people and vendors involved in each area of MEMS development: design, wafer fabrication, packaging, electronics, assembly, and so on. To halt this large machine while waiting for money to arrive is a huge waste of valuable momentum. People, knowledge, and technical capabilities will dissipate during downtime, and they may not be possible to immediately recapture when funding finally reappears.

In our business, we have seen our customers suffer from funding disruptions that lasted 6 months to more than a year. The end result was a profound waste of time and money while engineers' memories had to be refreshed, tooling rebuilt, and processes relearned. One of our customers was forced to suspend their foundry production for over a year due to lack of funds. When the customer finally received funding to resume production, the foundry rejected their order. So much time had elapsed that the foundry's business model and executive team had since changed. They viewed our customer as a sales prospect subject to their new criteria, not an existing customer deserving special consideration. Our customer was ultimately forced to seek a new foundry and to repeat all the foundry development stages, at great cost and time penalty to their product plans.

Summary

We have defined five stages of MEMS development in order to help orient those readers who might be new to MEMS product development. We have defined the stages according to the progress of the MEMS device development only. As will be explained in later chapters of the book, many other key components that accompany the MEMS product will also be developed during these stages, such as ASIC, packaging, and software, to name a few. We make reference to these five development stages throughout the book.

The stages of MEMS development are defined and gated according to challenges in MEMS design, fabrication, and test, and also by how the MEMS industry is structured given the use of foundries and other third parties for manufacturing of product. During the proof-of-concept prototype stage, you establish the motivation to pursue the development of your product. In the advanced prototypes stage, you iterate and refine your design in an environment that has the flexibility to improve your design and process flow. In most cases, the development then transfers to the foundry, which becomes your company's collaborator and partner in product development. In the foundry feasibility stage, you establish that your device can be made by your selected production foundry. In the foundry pilot production stage, the foundry verifies the process window for your device, and finally in the foundry production stage, your MEMS device is manufactured in volume.

Finally, planning and securing the funding required for all of these stages is as essential as the technical milestones you will achieve throughout the development process. Chapters 4–6 provide information that will help you to secure the right level of funding for successful product development.

References

1. McDonald CJ (1998) The evolution of Intel's Copy EXACTLY! technology transfer method. Intel SEMATECH
2. Huff M (2020) Process variations in microsystems manufacturing. Springer Nature, Switzerland

Part II
Business Requirements for a Viable Product

Chapter 4
What Is the Product? Requirements Analysis

Most innovators think ahead to the potential for a monetizable product while their technology is still in the form of a bare chip attached to a rack of electronics, and they are working on publishing results or demonstrating performance to potential early investors. The chip, or the invention, is not yet a product; it is a hopeful future product, a twinkle in the innovators' eyes.

The next step is to define: what is the product?

A product is a sellable item, something that will induce a person to part with their money. Creating a product therefore begins with the task of understanding which specific items, features, or functions will inspire a sale, and which will not. Defining the product begins with investigating and understanding the customer's requirements for the product. This chapter offers guidance on how to define product requirements for MEMS technologies.

Understanding the Market

Product requirements flow to and inform many other key steps in the formation and execution of a technology product business (Fig. 4.1). A wise entrepreneur, or product manager within an established company, uses an initial set of product requirements to help estimate the product's unit cost of manufacture, which then determines if the product could be sold at a profit at market prices. The quantity of potential profit derived from unit sales (the gross profit margin) ultimately determines if a business opportunity truly exists (Chap. 5).

The product requirements and the target unit cost together determine the scope of development work because development involves not only achieving functionality, but also achieving functionality at a unit cost that fulfills the business model. After a development plan has been appropriately scoped, then it will be possible to create a development plan cost model for bringing the product to market (Chap. 6).

© The Author(s), under exclusive license to Springer Nature Switzerland AG 2021
A. M. Fitzgerald et al., *MEMS Product Development*, Microsystems
and Nanosystems, https://doi.org/10.1007/978-3-030-61709-7_4

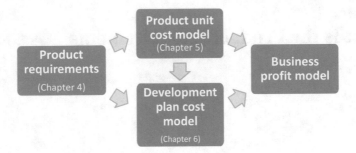

Fig. 4.1 Product requirements inform the unit cost model and the development plan cost model, which, in turn, inform the business profit model

Finally, the unit cost model (the calculations of the recurring costs of manufacturing) and the development plan cost model (the nonrecurring investment to create the product) inform the overall business model. Ultimately, profit from product sales must pay off the development costs within a reasonable time for the company to provide return on investment and continue as a sustainable enterprise.

Market Perspective Frames Product Requirements Analysis

To begin product requirements analysis, first consider whether your future product falls into the state of "technology push" or "market pull." These two terms describe the existing market's perspective on your product and provide a coarse categorization that frames how best to pursue product definition.

Technology Push

"Technology push" means that the customer market does not yet know it might want your product. Customers may not know that your technology exists or that it could be useful to them, or they might already be using some other existing technology to satisfy their application needs and not be looking for new options. Most new MEMS products, by virtue of being truly novel technologies, get pushed into the market by their makers.

In technology push, the main challenge will be to convince customers to try something new. If they do not know how or why to use your future product, you will need to first teach them about your product and how to implement it before they would even consider buying it. You may need to overcome the customer's skepticism that your new product could truly deliver what you promise. There might also be important considerations peripheral to implementing your new product, such as the customer's need to redesign a circuit board interface to accept your product or to retrain its employees.

If your new product is intended to displace an existing product, you will have to convince customers to give up the comfort and safety of the familiar product, and the tooling, processes, supply chain, and so on that go along with it, and replace it with your new product. The reluctance of customers to give up an existing solution, even if mediocre, for a superior new solution, often surprises entrepreneurs.

Human psychology and institutional inertia are also barriers to overcome in a technology push situation; convincing any human to give up what is familiar and safe for something new and unproven is always a difficult task. Humans generally only take chances on the unfamiliar when they perceive excellent benefits and minimal risks (this also applies to trying new foods, not just new technology products).

SiTime's silicon MEMS oscillators emerged to the marketplace in the mid-2000s, aimed at replacing the decades-old quartz crystal oscillator components that are found in nearly all electronic products. Despite the fact that its MEMS oscillators are smaller, higher performing, and consume less power than a quartz oscillator, market uptake was slow. Even when provided with convincing data of the silicon oscillator's superiority, SiTime's customers were initially reluctant to let go of the cheap and familiar quartz crystal [1].

In technology push, product requirements analysis therefore requires first developing a deep understanding of the features and cost of any competitor products, as well as any associated tooling and infrastructure used in assembly of the product into a larger system (such as an automobile or a smartphone). You must also understand other intangible or nontechnical issues, such as a customer's anxiety about disrupting a long-standing relationship with an existing supplier, or a competitor having a large catalog of components which enables a customer to enjoy "one stop shopping" and special deals on bundles of components.

This context is important to identifying the entire set of requirements for your product. Your product may need to be not just better, but ten times better—whether in performance, cost, or both—to entice the customer. Product design and development will proceed very differently if the product must be ten times better versus just incrementally better in order to be compelling to customers.

Market Pull

"Market pull" means that customers already have a known yet unfulfilled need for your product. In this situation, customers will be able to provide a detailed list of the features, performance, and cost they seek. It could even be that they have been waiting years for a product like yours to arrive. Your goals in this situation will be to discover their list of desired requirements and the product price they will tolerate and then to get the product made as quickly as possible. Chances are that your competitors are also alert to any market pull opportunities and are already racing to beat you to them.

In a market pull situation, customers may be so eager for the product that they will gladly place deposits or prepay while they wait patiently for your product to arrive. This is an ideal situation for an entrepreneur because the customers' money can help fund development. A recent example of a new product subject to extreme "market pull" is the Tesla Series 3 electric car; customer deposits for the car totaled hundreds of millions of US dollars, even though manufacturing was not yet up to full scale and wait times for delivery were over a year [2].

Market pull for several types of MEMS products in the United States occurred due to regulations issued by the National Highway Traffic Safety Administration (NHTSA). Federal Motor Vehicle Safety Standards (FMVSS) No. 126 and No. 208 required that all vehicles sold in the US be equipped with electronic stability control systems and airbag safety systems, respectively. When issued, these regulations created a strong market pull for MEMS accelerometers and gyroscopes beginning in the 2000s [3]. FMVSS No. 138 required tire pressure monitoring systems (TPMS) in all vehicles, which also increased demand for MEMS pressure sensors and accelerometers [4]. In China, the Ministry of Industry and Information Technology (MIIT) recently issued standard GB 26149 mandating TPMS systems for all new vehicles sold in China beginning in 2020. This government-mandated market pull should have a similar impact on demand for automotive MEMS sensors in China.

When in the state of market pull, product requirements analysis will require thorough investigation of customers' and/or regulatory requirements. Identify the range of potential customers for your product early, and then invest time interviewing those customers to determine their essential product requirements. Similarly, if selling your product into a regulated market, be sure to research regulatory documents in all of the countries in which your product might be sold, and invest time to correctly interpret those documents.

Finally, market pull creates an imperative to get the product to market as quickly as possible while customer interest is still strong and before your competitors arrive; this means that manufacturability is an essential and priority product requirement. During product development, design choices should take advantage of established MEMS manufacturing methods to enable quicker ramp-up to production and fastest time to market.

In summary, before beginning product requirements analysis, make sure you have a good sense for whether you will be in a state of "technology push" versus "market pull." In the former, you must be sure to examine competitors' products that your product seeks to displace. If you are pushing a novel product with no existing competition, you must be sure to understand customers' fears and concerns about adopting new technology. Should you be lucky enough to be in a "market pull" situation, it is still important to understand the customers' wish list and top priorities, as well as to prioritize speed to market as a key product requirement.

Which Parties Dictate the Product Requirements?

Complex technical products often have a chain of several intermediaries between the technology development company and the original equipment manufacturer (OEM), the entity that sells the completely finished product for use. If your company is going to sell a MEMS or microsystem component, then your company's customers will likely be intermediary integrators that combine more components with yours in order to form a subsystem or module, which is then sold to the next entity in the supply chain.

An example of this is the automotive industry, in which MEMS sensor companies sell their components to Tier 1 companies, such as Continental or Delphi, which then integrate the MEMS into electronic assemblies, which are then sold to the automobile makers (OEMs), such as Volkswagen or General Motors. The customer who actually buys the completely finished product, in this example, an automobile, is the "end user," the last entity in the chain who actually employs the product in its intended function. This type of supply chain is also common in the consumer electronics and medical device industries.

Your company's product must always satisfy the requirements of the end user; if you will be selling a component, then you may have additional requirements imposed by any or all of the intermediary integrators between your company and the end user (see Fig. 4.2).

For example, if your component will ultimately be placed into an automobile, you will have environmental requirements imposed by the end user's potential driving conditions, which might range from −40 to +50 °C. In addition, you may have interface requirements, such as I^2C data protocol, imposed by an intermediary integrator, so that your component will be able to communicate with the integrator's electronics module once assembled.

Be sure to identify all the parties between your company and the end user that might impose requirements on your products. You must understand both the technical and business requirements of all those parties, and beware that intermediary integrators may sometimes impose tougher requirements than end users.

If you might be selling your product into multiple unrelated markets, be sure to understand the full supply chain and requirements of each of those markets. For example, a MEMS pressure sensor could be sold for use in automobiles as well as

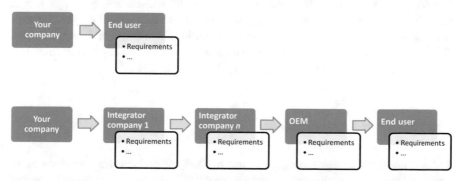

Fig. 4.2 If your company will be selling its product directly to the end user, you need only investigate their requirements (top). If your company will be selling a component that will be integrated into a more complex product, you will additionally need to investigate the requirements of all integrators (1 to n) and the OEM between your company and the end user (bottom)

in medical equipment. Each of those industries has their own supply chains, methods of doing business, regulatory requirements, and end user requirements to be investigated.

Understanding Technical Requirements

There are five key sets of technical product requirements to investigate:

- Functional.
- Environmental.
- Interface and package.
- Calibration.
- Quality and regulatory.

Early in development, it is easy to fixate on the device's functional requirements because the daily work of the engineering team in the prototyping stages will be getting the device to function well. During product requirements analysis, however, it is absolutely essential to investigate, understand, and address the other four sets of requirements in parallel because these are also essential to the product's success. No customer will purchase a product that performs beautifully on a laboratory benchtop but becomes inaccurate when exposed to the end user's operating temperature range, for example. The product must be engineered to fulfill *all* requirements. The earlier you can consider all five key requirement sets and incorporate them into your product development plan, the better.

Functional Requirements

For a sensor, the basic performance parameters to investigate will be power, range, sensitivity, linearity, and accuracy [5]. For an actuator, they would be power, range (of force, displacement, and so on), resolution, linearity, and repeatability. Depending on the end application, there may be many more functional requirements to fulfill. Some common parameters are outlined in Table 4.1.

Environmental Requirements

The end use environment strongly informs MEMS, electronics, and package design choices. Depending on the application, a number of different environmental factors may impact the MEMS product simultaneously. To become a viable product, the MEMS must not only survive in these environments but be able to perform in them consistently over time.

Table 4.1 Common performance parameters for MEMS sensors and actuators

MEMS device type	Common performance parameters
Sensor	Range Sensitivity Linearity Offset Accuracy Resolution Current draw/power consumption Selectivity Frequency response Noise Drift
Actuator	Range (force, displacement, etc.) Resolution Linearity Accuracy Current draw/power consumption Frequency response Noise Drift Hysteresis
Passive mechanical structure (micro-mold, microfluidic channel)	Dimensional tolerances Surface material Surface roughness Thermal conductivity Thermal emissivity Hydrophobicity/hydrophilicity
Passive electrical structure (through silicon via)	Dimensional tolerances Electrical resistance Electrical isolation Capacitance Parasitic capacitance Breakdown voltage Thermal conductivity Maximum power
Passive optical structure (grating, micro-mirror)	Reflectivity (over specific wavelength) Transmissivity (over specific wavelength) Thermal conductivity Maximum power

From a cost perspective, understanding the end use environment informs not only the contribution of environmental protection features to unit cost, but also the expected cost of testing to demonstrate survivability and lifetime. This cost of testing is incurred during development as well as during quality control, burn-in, or proof testing prior to product shipment (See Chap. 20 for more information on manufacturing tests.)

Table 4.2 Environmental factors, in order from more common to less common, that drive both feature and test requirements in certain end use applications

Environmental factor	Relevant to end use applications (examples)
Temperature	All
Vibration	Aerospace, automotive, consumer electronics, industrial
Impact/shock	Aerospace, automotive, consumer electronics, industrial
Humidity	Aerospace, automotive, consumer electronics, industrial, medical
Waterproofing	Aerospace, automotive, consumer electronics, industrial, medical
Chemical—Gas	Automotive, industrial, medical
Chemical—Liquid	Automotive, industrial, medical
Particles	Agriculture, automotive, industrial
Biological	Agriculture, consumer electronics, diagnostics, medical devices
Ionizing radiation	Aircraft, medical equipment, military, spacecraft
Light intensity	Agriculture, industrial, optical telecom

For high-volume products destined for consumer electronics or automotive applications, proof testing may occur on a statistical sampling basis—that is, some small percentage of a manufacturing batch would be subjected to testing. For low-volume products going to specialty applications, such as medical devices or aerospace vehicles, 100% testing of each individual unit in the environment may be required, thereby adding significant unit cost. You should understand what level of testing, and its respective unit cost, will be required for every potential end market.

Table 4.2 outlines common environmental factors that impose requirements on MEMS design, packaging, and electronics by relevant application.

Interface and Package Requirements

Defining interface requirements between the MEMS device and the next level of integration requires an understanding of the subsystem or product the MEMS device will be installed into and how it will interact with its surroundings. At minimum, the physical volume, mechanical interconnect, electrical interconnect, heat transfer, and any software requirements must be known.

Additional significant interface requirements include:

- Special mechanical enclosures or mounts, such as for inertial sensors.
- Custom electrical interfaces such as high density interconnects or shielding to minimize electromagnetic interference.
- Surface coatings such as for environmental protection, biocompatibility, or acoustic impedance matching.
- Specialty optical coatings such as for image sensors.
- Vibration damping.
- Chiller plates, in the case of systems with heat generation.
- Software drivers and algorithm support for end use application.

For MEMS products, the main interface between the MEMS device (chip) and the subsystem or printed circuit board (PCB) is the "package." (See Chap. 9 for more information on the components in a typical MEMS product.) Package is a generic term used to describe the enclosure in which a MEMS chip is mounted and where mechanical and electrical connections occur. The choice of package type depends on the end use operating conditions and environment, and different packages may have very different technical constraints and costs [6]. Package customization may be needed to enable the MEMS device to function as intended. It is also possible that no package might be needed if, for example, your MEMS component will be sold in the form of bare die to an intermediary integrator.

It is essential to understand which type of package will be required for the end use application because the package is usually a significant contributor (greater than 50%, in some cases) to overall product unit cost. Table 4.3 provides an overview of the most common MEMS package types.

Table 4.3 A sample of common MEMS products, their package requirements, type, and relative cost (Source: AMFitzgerald)

MEMS products (examples)	Package technical requirements	Package type	Relative cost of package
Wafer-level encapsulated MEMS	Through silicon via to enable electrical connection on exterior	Wafer-level encapsulation becomes the package	Low
Inertial sensors, resonators, RF filters	High-volume, low-cost packaging method for robust MEMS chips	Plastic overmold, such as quad flat no-leads (QFN)	Low
Microphones, microspeakers, pressure sensors	High-volume packaging method	Plastic molded case, with aperture	Low
Gas sensors, bolometers	Stiff package that tolerates higher temperatures, chemically resistant	Ceramic open cavity with lid	Medium
Micro-mirrors, electrostatic switches	Stiff package that tolerates higher temperatures; laser-welded lid to seal in nitrogen or argon atmosphere that provides gas damping	Ceramic open cavity with lid and controlled atmosphere	High
Medical, industrial or aerospace pressure sensors	Environmental isolation	Titanium, oil-filled	High
High precision resonators and sensors	Stiff package with heating unit and control loop to maintain stable temperature	Ceramic, ovenized	Very high
Sensors and actuators for radiation environments in aerospace, military, or medical applications	Specialty package, custom order	Ceramic, radiation hardened	Very high

At the time of writing, "Low" corresponds to less than US$0.10 per unit; "Medium" to less than US$1.00 per unit; "High" to less than US$10.00 per unit; and "Very high" to more than US$10.00 per unit

Calibration Requirements

Calibration is a critical aspect of sensor and actuator performance that is easy to overlook until late in the development process. It is, however, a key factor to the successful operation of a product in the field. Understanding the end use calibration requirements will crucially inform the MEMS device design, package design, electronics design, and algorithms for software.

For example, for a product to be used in a low-cost application, it will probably not be possible to include the cost of a microprocessor and memory chips in the sensor system in order to implement a complex polynomial fit calibration to correct a nonlinear sensor response. In this example, there will be a strong financial motivation to design the sensor to have a linear response so that its output may be calibrated and trimmed with a lower-cost solution, such as a simple variable resistor or analog circuit.

If the end use requires accurate sensor calibration to be maintained over years with no opportunity to recalibrate—for example, in implantable medical use—then excellent sensor stability is an important product requirement. During development work, designing the MEMS sensor to maintain that level of stability will be a priority. If it cannot be natively designed for stability, then it will require electronics and algorithms to compensate for drifting sensor performance over time. In this example, calibration requirements would strongly impact overall system design, product complexity, and product unit cost.

The inability to appropriately calibrate a sensor for its intended end use will result in an unsellable product. In contrast, the ability to maintain a stable and precise calibration over time makes a sensor quite valuable for high-performance applications and allows premium pricing of the product.

Quality and Regulatory Requirements

Most MEMS products will be sold into markets that have established quality standards, many of which have been formalized into documents listing specific performance requirements and testing methods. In the specific case of products for the medical markets, there will be regulatory requirements to meet and approvals to receive, such as CE mark in Europe or FDA approval in the United States, before it will be possible to sell the product.

It is vitally important to understand the quality and regulatory requirements of the end use application because these will inform all aspects of the development from an early stage. They may even dictate how the development work needs to be documented, in the case of FDA approval. A product's inability to meet these requirements will result in having to redo some or all of the development, driving up investment costs and delaying the timeline considerably.

Quality standards concerning acceptable rates for in-field failures, common in the automotive and aerospace industries, will demand extensive lifetime testing of the product before it can be sold. For example, an automotive manufacturer may demand statistical reliability data demonstrating a part-per-million failure rate before committing to making a purchase. Make sure that these requirements are known early in development. The MEMS device, package, and electronics within the product must be designed to accept and survive any reliability tests. The overall development plan must also factor in the time and resources needed to accomplish all required testing.

See Chap. 14 for more information on reliability testing during development and Chap. 20 for more information on product quality testing during production.

Understanding Business Requirements

In addition to technical requirements, products also have business requirements that must be fulfilled in order to sell them successfully. Typical business requirements are:

- Unit price.
- Overcoming existing barriers to entry.
- Satisfying intermediary integrators.
- Arranging security of supply.

Unit Price Requirements

If a new product will compete against and attempt to displace an existing product, the unit price to beat must be known at the start of product development, and will be a requirement that informs design choices along the way. Market data on existing competitors' products and knowledgeable market consultants can help to define the appropriate unit price requirement for your product.

If a product will be completely new and without initial competition, then establishing a unit price is a more difficult task. Market research should be conducted throughout development to continually explore and assess your potential customers' price sensitivities.

You must also forecast how unit price may evolve several years into the future. Today's market data is based on historical prices from the past few years, and your product may not be ready for sale until several more years in the future. Most technology product unit prices tend to decrease with time as competitors arrive and commoditization begins—the phenomenon known as price erosion. Price erosion rates vary among MEMS devices and even among different markets for the same type of MEMS device. For example, commodity MEMS inertial sensors for smartphones have experienced severe price erosion; however, high-performance inertial sensors have not [7].

Generally, higher-performing devices will command higher unit prices. However, due to the additional quality expectations of high-performance devices, described above, high prices do not necessarily imply high profit margin.

It is important to have at least a basic understanding of product unit price requirement early in your product development. In Chap. 5, we will describe how to estimate the unit cost of manufacturing of a potential product, even at an early stage of development. The difference between the two is the gross profit, which funds a company's fixed operating expenses. A company with insufficient gross profit margin will eventually bankrupt.

It is crucial throughout development to continually monitor the potential gross profit margin of your product and to actively manage the spread between the price the market would bear and the evolving unit cost of making it. Without active control during development, well-intentioned technical changes might cause the unit cost of a product to creep high enough to exceed current market prices, which would render the product unsellable and your business model kaput.

Overcoming Existing Barriers to Entry

When attempting to displace an existing component used in a product as complex as a smartphone or an automobile, beware that OEMs will not absorb the time nor the risk to redesign their complex product around your one new MEMS component. Your MEMS component product therefore might initially have to match the footprint and interface of an existing component's socket (hence the semiconductor industry vernacular, "winning the socket"). Your product will be perceived as more compelling and less risky if it can conform to the existing socket, even if your technology could enable a smaller size. A key business requirement in this situation is to understand all the requirements associated with the existing socket.

Eventually, when the customer reaches the point of considering a next-generation product redesign, and they would like to keep using your product, they might change their design to leverage your product's features. Achieving this point is a business victory because once your product achieves "design-in," as it is called, you have created a very effective barrier against the next competitor who might arrive.

When wishing to displace a competitor's component product from a system product, the burden will be on your company to prove to an OEM that your new MEMS component product will not change the function of the system product, nor increase assembly costs, nor introduce any risk of field failures.

When analyzing business requirements, it is essential to establish the criteria that will beat an incumbent. Would your MEMS component need to be three times cheaper or five times cheaper than the existing one? It is important to investigate and understand what tipping point will get an OEM or customer to take a chance on your product.

Meanwhile, if the existing supplier were to catch wind of your company's desire to displace their component from the OEM's product, then while you are attempting to prove yourself worthy to the OEM, the competitor may start to fight back. Common defensive tactics are discounting prices or increasing product quality to persuade the OEM to keep using their component. Depending on the competitor's reaction, your business requirements may become a moving target as the competitor continually adjusts its strategy. An incumbent offering an aggressive price discount could potentially destroy your hoped-for gross profit margin as well as cause a change in the OEM's risk versus reward calculus on accepting your new product.

Satisfying Intermediary Integrators

As described earlier, in some cases your company's immediate customer may be an intermediary integrator rather than the end user of your product. When an intermediary is involved, the engineering and business challenges multiply due to all the parties involved or influencing the purchase decision. Be sure that you investigate and understand the business requirements of all parties.

When dealing with an intermediary integrator, an additional human factor to consider is that they may have a reputation or history with the OEM (or the next-level intermediary integrator) which they seek to protect. If they perceive your new product as posing a risk of disrupting their existing business, whether through production delays or unanticipated costs or field failures, then they will be quite reluctant to integrate your product. Understanding how to address and allay their fears is a crucial business requirement.

Arranging Security of Supply

For products heading into high-volume markets (greater than 100 million units per year), intermediary integrators or the OEM company purchasing your product may impose the business requirement that your company have a robust supply chain before they would consider buying your product. This could mean that they would expect all of your suppliers to be well-established companies with existing manufacturing infrastructure and proven ability to supply that annual volume of units.

For example, the intermediary integrator or OEM customer might demand production on 200- or 300-mm wafers, or use of package types that enable a high degree of assembly automation. These business requirements influence engineering requirements and vendor choices so therefore need to be considered early in MEMS product development. Continuing this example, production on 200- or 300-mm wafers implies (at the time of writing) that wafer fabrication must employ only processes available at large silicon foundries and would therefore rule out use of materials or methods that might otherwise be feasible at smaller-volume foundries (such as use of gold or xenon difluoride release etch).

Your company's suppliers might have to demonstrate that they have excess capacity or an ability to respond quickly to spikes in demand. If product demand were to surge, such as during end-of-year holiday orders, there should be no gap in inventory.

Finally, an intermediary integrator or OEM customer may also demand that your company have a fully qualified second or third production facility to mitigate risk of supply disruption. Qualifying two or more production facilities is a significant undertaking, so that time and budget must also be factored into the overall product development plan.

Summary

After all this research into product requirements, you may discover that what you initially thought would be the product, is not the product that customers actually want to buy. As jarring as that discovery might be, it is a golden insight for product development.

Knowing your customers' desires and, therefore, your product's true technical and business requirements, in the early stages of MEMS product development, will save time, money, and many headaches over a multiyear effort. Talking to prospective customers, intermediary integrators, and OEMs on a regular basis before and throughout development is the best way to gather this information, as well as to plant seeds for future sales.

Venture capital-backed companies working on short timelines and limited budgets especially need to be diligent about discovering and meeting essential product requirements and resisting the addition of nonessential features. "Feature creep" eats precious time and resources and delays product launch, which can fatally delay company revenue. A product might be brilliant, but if it cannot be sold at a high enough price to generate sufficient gross profit margin to operate the company and pay back the investment in product development, it will eventually bankrupt a young company.

Investing time in uncovering and prioritizing all of your product requirements while still early in development will pay dividends by defining a stronger, more desirable product. It will additionally help scope a more accurate timeline and budget for the product development, which in turn, will create a more realistic business model and increase the product's ability to successfully reach the marketplace.

References

1. Petersen K (2014) "MEMS Entrepreneurial Perspectives," Solid-State Sensors, Actuators and Microsystems Workshop Hilton Head Island, South Carolina, pp. 64–67
2. Vlasic B "Tesla's Model 3 Already Has 325,000 Preorders," *The New York Times*, April 7, 2016
3. "MEMS price to fall as production rises," *DesignNews*, September 22, 2008. (https://www.designnews.com/automotive-0/mems-price-fall-production-rises/169648056249814)
4. M. Löhndorf, T. Lange, "MEMS for automotive tire pressure monitoring systems," Chapter in: MEMS for automotive and aerospace applications, Eds. Michael Kraft, Neil M. White, Woodhead Publishing, 2013, Pages 54–77
5. Fraden J (1993) Sensor characteristics. In: AIP handbook of modern sensors: physics, designs and applications. AIP Press, New York
6. Lau JH, Lee CK, Premachandran CS, Aibin Y (2010) Advanced MEMS packaging. McGraw-Hill, New York
7. Sperling E "The trouble with MEMS," *Semiconductor Engineering,* May 25, 2016. (https://semiengineering.com/the-trouble-with-mems/)

Chapter 5
Is There a Business Opportunity? Product Unit Cost Modeling

With the guidance of Chap. 4 to explore product requirements and to define what your product will be, this chapter will help to answer an essential question: is there really a business opportunity for your product?

What we mean by this is whether it appears possible to make a profit by selling your product, and whether that profit might eventually be large enough to both pay back the development investment and to sustain ongoing business operations. The first step towards answering the question of business opportunity is to build a unit cost model for your product.

Building a unit cost model is often overlooked in early product development. You might think it is simply not possible to build a useful estimate at such an early stage and disregard the task. With so many aspects of a product design unfinished and vendors not yet chosen, estimating manufacturing unit cost may feel premature. Excitement about a new technology and the promise it holds for commercial use may eclipse mundane matters like profit margin. Yet as this chapter shows, one benefit of the relatively mature industry of wafer-based manufacturing is that there are rules of thumb that make it simple to create quick and useful cost estimates.

Ideally, an unit cost model should be created the moment you begin to think about developing a product. It should be analyzed well before any prototypes are built, not during, nor after!

Rough Order of Magnitude Unit Cost

The first step is to create a simple model of the rough order of magnitude (ROM) unit cost. If this estimate looks reasonable, it will be worthwhile to proceed with initial development. As understanding of the design and its manufacturing advances throughout product development, you will be continually updating and refining the unit cost model.

© The Author(s), under exclusive license to Springer Nature Switzerland AG 2021
A. M. Fitzgerald et al., *MEMS Product Development*, Microsystems
and Nanosystems, https://doi.org/10.1007/978-3-030-61709-7_5

If the initial ROM unit cost were to show poor profit opportunity, then it is time to take a step back and revisit the product requirements before continuing any product development. It is especially important to do market research on any products to be sold to the consumer or automotive markets, where the average selling prices (ASP) of MEMS products are quite low (<US$0.50 per unit). These markets have many entrenched, large competitors that could beat your product simply by lowering their prices.

For example, if you estimate your MEMS product's ROM unit cost at US$100 per unit, and the product will be sold into a market that would pay only US$0.10 per unit, at three orders of magnitude apart, frankly, there is no chance of a business opportunity. Not even the economy of scale of volume manufacturing could close a gap that large (more on that later).

However, if the ROM unit cost were US$0.01 per unit, and the market will pay US$0.10 per unit, at a potential gross profit margin of 90%, now that is a very interesting business opportunity indeed. There is plenty of margin, so even if feature creep later swells your unit cost to US$0.03 per unit, there would still be a healthy 70% gross profit margin available.

So how could you possibly estimate unit cost when you have not even finalized the entire fabrication process or packaging needed to build the MEMS product? Fortunately, even without that information you can calculate a ROM unit cost for a MEMS product in two steps using only four estimated numbers and one MEMS industry rule of thumb:

- Expected dimensions of the MEMS die, in millimeters.
- Usable wafer area, based on wafer diameter.
- Total number of units expected to be sold in a year.
- Approximate cost of a MEMS production wafer for the volume of wafers to be purchased.
- Rule of thumb that 75% of a MEMS product's total cost is due to packaging, electronics, and calibration, or, conversely, that the bare MEMS die comprises only 25% of total unit cost. (Refer to Chap. 9 for more discussion on the components of a MEMS product.)

Step 1: Calculate Number of Wafers Needed to Fabricate the Number of Units to be Sold in a Year

Assume your MEMS will be produced on a 200-mm-diameter silicon wafer, which has approximately 30,000 sq. mm of usable area.[1]

---

[1] The usable area provided is suitable for rough estimates only. The usable area of a silicon wafer is ultimately determined by the wafer perimeter keep-out zone and area consumed by lithography alignment marks, test structures, and dicing streets.

To calculate the area of the MEMS die (Eq. 5.1), first add 0.1 mm to both the X and Y chip dimensions to account for the dicing streets. If using stealth dicing or any zero-kerf separation method, it is not necessary to include this dimension. If the chip will be less than 1 mm long in its X or Y dimension, it is especially important to include an accurate dicing street dimension because the street will consume a significant fraction of the wafer's area.

$$\text{Die area (in mm, including dicing street)} = (X + 0.1) * (Y + 0.1) \qquad (5.1)$$

Next, estimate the number of die (or units) per 200-mm wafer; for example, if the die area is 6 sq. mm, including dicing street, calculate as follows:

$$\text{Estimated units per wafer} = 30,000 \text{ sq.mm / die area} \qquad (5.2)$$

$$= 30,000 / 6$$

$$\text{Estimated units per wafer} = 5,000 \text{ units at } 100\% \text{ yield}$$

The units per wafer estimate in Eq. 5.2 is approximate because, in practice, die layout (arrangement on the wafer) and the need for alignment marks or test areas will reduce the overall number of units per wafer. There are many calculators available on the internet that can more accurately estimate the number of die per wafer, accounting for die shape and wafer layout geometry.

With the estimated units per wafer, we can now calculate the minimum number of wafers needed per year. To create 100 million units per year, for example:

$$\text{min. \# of wafers} = \text{\# units to sell annually} / (\text{est.units per wafer} * \text{yield}) \quad (5.3)$$

$$\text{min. \# of wafers} = 100,000,000 \text{ units per year} / (5,000 \text{ units per wafer} * 100\% \text{ yield})$$

$$\text{min. \# of wafers} = 20,000 \text{ wafers per year}$$

Note that MEMS wafers rarely yield at 100%, so Eq. 5.3 calculates only the minimum number of wafers per year. For now, we are keeping the math simple. More realistic numbers for yield will be discussed in the section on die unit cost below.

In MEMS manufacturing, the production cost per wafer varies with the number of wafers produced in a year due to the economy of scale. Table 5.1 provides some

Table 5.1 Approximate cost of a 200-mm MEMS production wafer for a medium-complexity MEMS process flow as a function of the order of magnitude volume of wafers purchased per year, not including dicing or testing

Minimum annual volume of wafers	Approximate cost per wafer
100	US$4,000
1,000	US$2,000
10,000	US$1,000

rough numbers, valid at the time of this writing, for the cost of a MEMS production wafer, scaled by the number of wafers to be purchased in a year.

Wafer cost also depends strongly on the type of substrate material and the process flow; however, for the purposes of a ROM unit cost, the numbers in Table 5.1 are sufficiently accurate.

Step 2: Calculate ROM MEMS Die Unit Cost

You can calculate a ROM MEMS die unit cost using the wafer cost from Table 5.1. Continuing this example, select the wafer cost based on a minimum volume of 10,000 wafers per year:

$$\text{ROM MEMS die unit cost} = \text{wafer cost} / (\text{est. units per wafer} * \text{yield}) \quad (5.4)$$

$$\text{ROM MEMS die unit cost} = \text{US\$1000} / (5000 * 100\%)$$

$$\text{ROM MEMS die unit cost} = \text{US\$0.20 per unit}$$

With the estimate of the bare MEMS die cost, using our rule of thumb that the MEMS die comprises only 25% of the finished sensor price, we conclude:

$$\text{ROM finished unit cost} = \text{US\$0.20} * 4 = \text{US\$0.80 per unit} \quad (5.5)$$

If the end use market could tolerate a unit price of US\$1.00, for example, a ROM unit cost of US\$0.80 means there could be an opportunity to generate US\$0.20 of profit per unit sold.

Keep in mind that this estimate is rough because, among other things, it assumes 100% die yield, which is not a realistic condition. We will refine the model further below. For the purposes of an initial check, however, this result provides an existence proof for the potential of profit.

Gain Insights from Parameter Sensitivity Analyses

Early-stage MEMS development does not always proceed as planned. Now assume that after the packaging and test engineers have their input and insist on adding a few more bond pads, and then the marketing team demands that a company logo be added, the die grows to 12 sq. mm in area, including dicing street. If we rerun the ROM calculation using Eqs. 5.2, 5.4 and 5.5, the ROM unit cost will be US\$1.60 per unit, or US\$0.60 over the price target. Our simple example assumed a large annual volume of wafers, so the production wafer cost is already quite low; we also optimistically assumed 100% yield. It is clear from this example that if the die were to double in size, they could not be sold profitably.

When creating cost models, even simple ones, it is important to analyze what-if scenarios to determine the model's sensitivity to certain key assumptions or parameters. In this example, we examined only the effect of die area. Variations on expected yield and annual sales volume also need to be analyzed. For example, if the annual product volume estimate proved to be too optimistic, resulting in far fewer wafers purchased per year, the per wafer cost would be higher than expected, and the unit cost would increase accordingly.

One major insight from this simple example is that die size is a very important product specification that must be carefully controlled throughout the MEMS development. Business people readily understand the concept of estimating and then controlling manufacturing costs to keep unit costs in line with the business model. However, the business impact of a creeping increase in die size throughout the course of a development might be easily overlooked. If, during MEMS development, any engineering discoveries suggest that die area needs to be increased, the unit cost model must be reevaluated immediately.

Detailed Unit Cost Model

To build a more refined unit cost model, you will need more information on the design of the MEMS and its fabrication process, as well as all the peripheral components that comprise the product. At a rudimentary level, you must understand the bill of materials (BOM) for the control and readout electronics; the package; and the assembly, test, and calibration conditions. Yield is also a very important variable in this calculation and should be reasonably estimated for each step of the manufacturing process.

This unit cost model captures the recurring expenses of ongoing manufacturing in three steps. The nonrecurring expenses of development and tooling will be described further in the next chapter.

Step 1: Identify all Operations or Components of the Finished MEMS Product and Expected Yield for each

To begin a detailed unit cost analysis, you must have enough of a concept design completed to know the overall operational flow as well as the main components required for the MEMS product to function in its end use application (Refer to Chaps. 12–14 for guidance).

Table 5.2 describes typical operations and purchased components used in the manufacturing of a MEMS product. This is an example; depending on the intended end use of your product, not all of these items may be relevant. The table also

Table 5.2 Typical manufacturing operations that contribute to MEMS product unit cost and the expected yield range for each operation. Assume that any purchased components (such as passive electronic components, package, or substrates) by default have a yield of 100%. (Source: AMFitzgerald)

Module	Manufacturing operation or purchased component	Expected yield at individual operation
MEMS fabrication	Starting wafer material	100%
	MEMS wafer process: Steps $1-n$	50–99%
	Wafer thinning	95%
	Electrical test	50–99%
	Dicing	95–100%
	Optical inspection	90–100%
Electronics	Readout chip (ASIC or ASSP)	100%
	Passive components	100%
	Package or substrate	100%
Assembly	MEMS die mount and interconnect	95–99%
	Readout die mount and interconnect	95–99%
	Environment: Fill gas or vacuum getter	90–99%
	Lidding or encapsulation	95–99%
Final test and calibration	Function test	50–99%
	Calibration	90–99%
	Final quality control test or burn-in	95–99%

describes the typical yields expected based on average manufacturing conditions common in industry practice at the time of writing.

Each operation listed in Table 5.2 may itself contain dozens of steps. For example, a fabrication process to create a wafer of MEMS devices may contain several hundred process steps, each of those having respective yields.

Table 5.2 highlights that there are a large number of operations in the manufacture and assembly of a MEMS product, and not all of them can or will yield at 100%. Pay close attention to the yields of the post-wafer fabrication steps (the "back end") that transform the bare MEMS die into a useful product. Because these operations occur individually to units, not in batches, yield loss at this late stage can be quite costly.

Of the 16 items in the Table 5.2 example, four are purchased components, so we do not need to worry about their yield, and 12 involve manufacturing operations. If we were to assume that the yield of each of the manufacturing operations were a respectable 95%, then compounding that yield over 12 successive operations, the overall yield of the entire manufacturing process would be:

$$\text{Overall yield} = (\text{yield of step 1}) * (\text{yield of step 2}) \ldots * (\text{yield of step } n) \quad (5.6)$$

$$\text{Overall yield} = (0.95)^{12} = 0.540 = 54.0\%$$

This simple calculation provides a crucial insight: even with 95% yield at each operation, the sheer number of operations results in a not-so-satisfying 54.0% yield overall. Revisiting Eq. 5.4, this new yield estimate will result in a unit cost of US $0.37, an 85% increase over the idealized 100% yield ROM unit cost estimate! Even slightly underestimating the impact of yield on your product unit cost can destroy your business model.

There are two "big levers" you control to improve the overall manufacturing yield and therefore to keep the unit cost as low as possible:

- Improve the yield of individual operations to even higher levels, such as 98%.
- Reduce the number of operations.

If it is possible to both increase individual operation yield and reduce the number of operations, that will of course provide the biggest improvements to yield.

Practically speaking, however, reducing the number of operations is usually the easier and less expensive way to boost overall yield. The optimal time to figure out how to reduce the number of operations is early in development, when it can often be accomplished through thoughtful design. Improving the yield of individual operations usually requires improvement in manufacturing methodology and/or better tooling or machinery, both of which require investment in time and money.

Step 2: Gather Cost Data for each Manufacturing Operation

Once each manufacturing operation has been identified, it is important to collect realistic cost data for each step. This task is actually harder and more time-consuming than it sounds. Gathering useful manufacturing cost data requires serious leg work: identifying potential vendors, signing non-disclosure agreements (NDAs) with them, sharing technical information about the operation they will perform for you, and finally soliciting a "budgetary quote" from them.

This budgetary quote is the valuable data your unit cost model needs. It is based on an initial assessment of your product's needs and the vendor's deep understanding of the manufacturing method to be applied. Most vendors are happy to provide these types of estimates as a professional courtesy with the understanding that the estimate is non-binding and subject to change. When a vendor provides a budgetary quote, they have made a quick cost estimate based on their initial understanding of your needs, but they have not yet done enough assessment of the exact operation requirements to provide a firm quote—that takes a larger time investment. Nonetheless, budgetary quotes are usually within ±15% of expected cost and therefore good enough for initial cost modeling. Be kind to your vendors and do not press them for more precise numbers until you are ready to place an order.

It may take a duration of weeks to months to gather this data from each vendor. Typically, this data gathering may proceed in parallel with multiple vendors. The time spent will be a worthwhile investment because it will also provide an opportunity to learn more about the details of the specific operation, the capabilities of the vendor, and your product's overall supply chain requirements.

You may also be able to quickly harvest some cost estimates from colleagues or published reports. These can be useful for initial models; however, it is important to make sure you understand the underlying assumptions of any third-party data. Manufacturing process costs are highly specific to both the operation itself and the volume of product moved through that operation. Published cost data derived from reverse-engineering analysis of high-volume products, such as accelerometers or microphones, may not be useful for modeling a specialty, low-volume product.

Finally, for each manufacturing operation, be sure that you understand the manufacturing unit and its batch size associated with the quoted costs. Most manufacturing operations are quoted as batch operations, not as cost per individual unit. For example, MEMS wafer foundries typically quote the cost for a finished wafer batch size of 23 or 24 wafers; wafers are processed in cassettes having slots to hold 25 wafers, and the foundry will typically reserve 1 or 2 of those slots for accompanying test wafers. In contrast, the batch size for a wafer grind and polish operation may be only five wafers because that is how many wafers fit on the polishing tool's platen. As you build your model to calculate unit cost, quoted vendor costs must be divided by the appropriate numbers to arrive at an accurate per unit cost.

Step 3: Assemble Model and Perform Parameter Sensitivity Analyses

With an understanding of the manufacturing flow, operations, and expected cost and yield of each, you may begin assembling your unit cost model. A spreadsheet is the simplest and most versatile tool for this job; however, MATLAB or other programming environments are also suitable. You will be revisiting your model and updating it repeatedly as the MEMS development progresses, so whichever tool you choose, make sure you and your colleagues are proficient users.

Table 5.3 outlines a simple model to calculate the unit cost of a MEMS product which consists of a MEMS and an ASIC wafer, bonded at the wafer-level and sold as bare die to an intermediary integrator. The inputs to the model are the total

Table 5.3 An example of a simple model to calculate a realistic unit cost of a MEMS product

Manufacturing operation	Operation cost per wafer	Expected wafer-level yield
MEMS fabrication	$ 2000	70%
Solder bumping	$ 50	99%
ASIC fabrication	$ 1200	98%
Wafer bond	$ 100	99%
Wafer thinning	$ 50	99%
Dicing, saw	$ 100	98%
Total	**$ 3500**	**65.2%**
Die per 200 mm wafer	**Total cost per die**	**Total cost per KGD**
10,000	**$ 0.35**	**$ 0.54**

number of die per wafer, the manufacturing steps, the cost of each step per wafer, and their respective, expected yields. From the yield of each operation, we can calculate the overall expected yield and therefore the cost of a "Known Good Die" (KGD). This provides a more realistic estimate of the cost of manufacturing a sellable unit.

As stated earlier, the real power of the model is that it provides the ability to perform parameter sensitivity analysis in order to instantly understand the impact of different scenarios to unit cost. If cost reductions might be required to bring the total unit cost into a specific range, the model may be used to identify whether reducing the cost of one or more steps will solve the problem. In particular, the impact of high cost, low yielding steps on total unit cost will be readily apparent in the model.

Insights gained from cost modeling should be fed back to the engineering and manufacturing teams. The model also serves as a communication tool; it can highlight the importance of key operations to product success, and unify the team so that they can sharply focus their collective efforts.

For example, if reaching the desired total unit cost requires the yielded cost of a certain step to decrease by 30%, three theoretical choices would be:

- Decrease the cost of the step by 30% while holding yield constant,
- Increase yield by 42.9% while holding cost constant, or
- Decrease cost by 15% while simultaneously boosting yield by 21.4%.

Consulting with the design team, the foundry and other vendor partners would help identify which of the three choices (or other combinations thereof) will be easiest to implement. Then the team should form a plan to make that choice reality.

Track your Assumptions and Iterate the Model when Assumptions Change

When working with a cost model, it is critically important to understand the explicit or implicit assumptions made within the model. Ideally, all assumptions should be documented so that any colleagues reviewing the model will have proper context of the cost calculations.

First and foremost, it should be clear to all what the cost model includes, what it does not include, which input data have been validated, and which are still estimates. For example, if initially you assume that you will use a PCB-mounted discrete electronic circuit, but then later decide to switch to a custom ASIC, those two items have very different costs. The impact of that switch must be reassessed immediately. A more subtle change to the cost model might result, for example, from switching wafer polishing vendors, if the new vendor's platen holds more wafers than the prior vendor, for the same batch fee. As MEMS development proceeds and more information comes to light, the model should be continuously iterated and expanded as needed.

Advanced Unit Cost Modeling

This chapter introduced the basics of cost modeling. Of course, it is possible to make models that are far more sophisticated than the simple ones presented here [1–3]. Our general advice is to begin with a simple model, identify and understand the "big lever" cost drivers, and address them. Then iterate the model and add complexity as new information is discovered.

An advanced cost model could incorporate some of the following items:

- The cost of borrowed money.
- Foreign exchange float, particularly if working with many foreign vendors.
- Amortized costs of special tooling, equipment, or facilities.
- Royalty payments.
- Shipping and insurance costs to move WIP between vendors.
- And more.

As product development continues and the cost model becomes more sophisticated, it can also provide valuable insights for business decisions such as financing, discounting strategies for winning customers, and vendor negotiations.

Beware of Common Blunders

Cost models can go quickly awry with inappropriate assumptions or input data. As the old critique about modeling elucidates, "Garbage in, garbage out."

Common blunders in MEMS cost models usually arise from unfamiliarity with MEMS manufacturing costs, faulty assumptions, or inappropriate use of data from another industry, such as semiconductor wafer production or metal machining. This latter blunder typically occurs when an entrepreneur may have significant expertise in the end use application or product development, but is completely new to MEMS manufacturing.

Another common error is to incorporate data from commodity, high-volume products or from high-volume semiconductor foundries when your product's volumes will be much lower or the process will be quite different. It may be true that a finished 200-mm semiconductor wafer could be purchased for US$600 from a high-volume fab in Asia, but this cost data is not at all useful to your model unless you plan to purchase at least 10,000 wafers per year from that fab at comparable process complexity.

Faulty comparisons frequently occur between semiconductor and MEMS process costs, which are more technological cousins than twins. For example, while CMOS processes may involve more than 30 mask layers, and your MEMS process may have only 6 mask layers, that does not mean you could expect your MEMS wafers to cost the same or less than a CMOS wafer. Although there may be many more steps in a CMOS process, many of those steps are executed in large batch processes using standardized and well-characterized process recipes, so the cost per step per wafer is quite low. Many MEMS process steps, notably DRIE and wafer bonding, are single-wafer, serial processes, with lower yield rates than CMOS process steps. For this reason, as well as lower overall wafer volumes, MEMS wafers are almost always more expensive than CMOS wafers, despite having fewer mask layers or fewer process steps.

Summary

A product unit cost model is a tool that connects the technical and business domains of a product development and provides a conduit for passing important requirements between them. Early in development is the optimal time to make architecture changes that could reduce product unit cost and improve yield.

Avoid naïve thoughts that everything will magically get cheaper in volume production. Economy of scale confers many benefits, but it will not make a complex and expensive product suddenly an order of magnitude or two cheaper to manufacture. Talk to your vendors to better understand how their processes scale and where advantageous break points in cost occur as volume increases.

References

1. Lawes R (ed) (2013) MEMS cost analysis. Jenny Stanford Publishing, New York
2. Mislick GK, Nussbaum DA (2015) Cost estimation: methods and tools. John Wiley & Sons, Hoboken, NJ
3. Ross R, Atchison N (2000) Yield Modeling. In: Nishi Y, Doering R (eds) Handbook of semiconductor manufacturing technology. Marcel Dekker, New York

Chapter 6
What Is the Budget for Development?

The development plan cost model determines the funding (nonrecurring expenses) needed to bring your technology from a pre-product state to market based on the product requirements identified in Chap. 4. The development plan cost model also estimates the cash flow needed over time and during different stages of the development. Investors and executives will not give a product development team a giant pile of cash for the entire effort; instead, to maintain some control over spending, they dole funding out in pieces, also known as tranches. Usually, the team needs to carefully manage its cash flow and then successfully complete a milestone in order to receive the next round of funding. This development plan cost model can, therefore, also inform overall company financial management.

Investors may provide initial funding based on the potential impact of your new product, but they will soon want to know when and what will be their return on investment. If you had, for example, told your investors that it would only take one year to get to the foundry production stage and then you were still iterating in the advanced prototypes stage one and a half years later, they may begin to doubt your ability to succeed. Knowing and communicating what it will realistically take to reach that point will also help to reduce the risk of investors pulling their support prematurely if your product were not immediately meeting expectations.

Understanding the Development Plan Cost Model Assumptions and Limitations

The method outlined in this chapter will estimate the budget needed to develop only the MEMS product. It will not examine sales and marketing expenses, company administration (overhead expenses), or other indirect costs of running a business. For a start-up company, the development cost model will be crucial to informing the appropriate amount of money to raise at each phase of the company's evolution.

© The Author(s), under exclusive license to Springer Nature Switzerland AG 2021   59
A. M. Fitzgerald et al., *MEMS Product Development*, Microsystems
and Nanosystems, https://doi.org/10.1007/978-3-030-61709-7_6

Before starting work on the model, it is important to define the endpoint of the development effort. You will likely need the consensus of both your colleagues and your investors or executives to define the endpoint. A typical endpoint of a development cost model is having the product ready for commercial sale—in other words, the point at which the product unit cost model begins. For a start-up company, the model's endpoint may be achieving a short-term milestone demanded by investors. In that case, once you reach that particular milestone, you would need to create a new cost model for the next milestone in your product development. No matter what the endpoint may be, it is important to clearly define it. The sum of money estimated by the model will depend on the endpoint selected.

To create a model, we employ some common industry estimating methods and rules of thumb, which were accurate at time of publication. You will need to make adjustments for your own local costs, salaries, and economic conditions of the time. Build the model parametrically so that you can easily customize and tweak the inputs to your specific situation.

Cost models may be as simplistic or as complex as you prefer or need. No matter the level of complexity, it is crucial to always understand and explicitly state the assumptions that underpin the model. Remain vigilant over time for any changing circumstances that may render those initial assumptions invalid and therefore corrupt the model's results.

In general, we recommend beginning with a simple model in order to understand the first-order relationships between inputs and outputs ("the big levers") before adding more detail and complexity. A very useful model can predict within ±20%; adding more detail in order to achieve ±5% accuracy usually is not worth the effort and can make the model fragile if the input assumptions become too specific.

Building a Level-of-Effort Cost Estimate

When estimating something as large and ill-defined as a multiyear development project, the most practical approach is to create a level-of-effort estimate instead of a task-based estimate. The level-of-effort estimate is based on how many people, over a certain duration, will be needed to accomplish a goal. Experience and engineering judgment are required to reasonably estimate the number of people and time needed.

In contrast, task-based estimating focuses on first identifying all the specific tasks to be achieved, itemizes the cost for each task, and sums up the costs to arrive at a total budget. When entering product development, it is difficult to predict the exact list and sequence of tasks that will occur over multiple years (especially those that will occur in the later years), so task-based estimating becomes too cumbersome.

Another challenge in building a MEMS development cost model from a list of tasks is not knowing how many design or product engineering iterations may be necessary to reach a product launch. By using a level-of-effort analysis, you will better understand the minimum funding that will be required at first, and then later, as the tasks are formed, you may estimate how many redesigns, parallel prototyping paths, and so on, your existing funding can support. For controlled project spending, the available budget must be an input into engineering decisions, not the other way around.

In Chap. 3, we discussed the main stages of MEMS product development and provided minimum order of magnitude cost estimates for those different stages. We will use those same definitions here, looking more closely at the proof-of-concept prototype, advanced prototypes, foundry feasibility, and foundry pilot production stages.

In Table 6.1, we illustrate how to build a parametric level-of-effort estimate for different stages of development. We group expenses into four categories: development team personnel, MEMS fabrication, CMOS fabrication, and vendor services. The first category is explicitly a labor cost, whereas the latter three are primarily material and services costs, although they do include some cost of labor. These four categories include only the engineering development costs and do not include other concurrent business operating expenses such as overhead, accounting, business development, or capital expenses (for example, equipment, software). The data are based on our direct experiences developing MEMS products, and the costs reflect execution by an experienced MEMS development team on 150 mm diameter wafers.

In the Table 6.1 Typical Quantity column, we provide ranges for the typical number of personnel and iterations each stage requires, based on our experiences developing MEMS products. If your MEMS device is relatively simple, then you might base your budget estimate on a smaller number of people and iterations. If your device is more complicated, you should assume that you will need more iterations at each stage.

Chapter 8 outlines the composition of the team you will need during the different stages of development. We recommend that you refer to it as a guideline to estimate headcount, an important input for the level-of-effort cost model. The responsibilities for the team will include MEMS and electronics design (modeling, layout, process integration), testing, data analysis, software development, and reliability and quality testing.

For estimating Typical Quantities of MEMS fabrication, you need to have an idea of how many unique fab runs can be realistically achieved in a year. In general, a medium-complexity MEMS device will require 2–4 months per wafer run, not including time spent waiting for specialty start materials such as custom silicon on insulator (SOI) wafers. Therefore a maximum of three to six runs per year will be likely.

Typical Quantities of CMOS fabrication assume that a design will be created and run on a standard CMOS foundry process where the desired number of wafers can simply be purchased without any development. Typical Quantities of vendor services include steps such as packaging, assembly, and specialty process services.

In Table 6.1, the Average Cost/Unit column is based on 2020 Silicon Valley rates. Costs may vary based on your location and the location of your vendors, and depending on the level of expertise. For example, a vendor that has developed custom packaging for similar MEMS devices may be able to capitalize on that

Table 6.1 A level-of-effort development cost model for a medium-complexity MEMS device being developed by an experienced team on 150 mm wafers. Table based on 2020 Silicon Valley rates (US$); when applying this model, update numbers for geographic location, device application, and level of complexity.

Development stage	Typical quantity	Average cost/unit	Duration range	Cost range
Proof of concept				
Development team	2 people	$150 k/year/person	0.5–1.5 years	$150 k–$450 k
MEMS fabrication	2 runs/year	$100 k/run	0.5–1.5 years	$100 k–300 k
CMOS fabrication	2 runs/year	$50 k/run	0.5–1.5 years	$50 k–150 k
Vendor services	3 vendors/year	$3 k/vendor	0.5–1.5 years	$5 k–$14 k
Subtotal (rounded up to nearest $50 k)				**$350 k–$950 k**
Advanced prototypes				
Development team	4 people	$150 k/year/person	1–1.5 years	$600 k–$900 k
MEMS fabrication	3 runs/year	$200 k-$400 k/run	1–1.5 years	$600 k–1800 k
CMOS fabrication	2 runs/year	$50 k/run	1–1.5 years	$100 k–150 k
Vendor services	3 vendors/year	$5 k/vendor	1–1.5 years	$15 k–$23 k
Subtotal (rounded up to nearest $100 k)				**$1.4 M–$2.9 M**
Foundry feasibility				
Development team	4 people	$150 k/year/person	0.5–1 year	$300 k–$600 k
MEMS fabrication	3 runs/year	$200 k-$400 k/run	0.5–1 year	$300 k–1200 k
CMOS fabrication	2 runs/year	$50 k/run	0.5–1 year	$50 k–100 k
Vendor services	3 vendors	$10 k/vendor	0.5–1 year	$15 k–$30 k
Subtotal (rounded up to nearest $100 k)				**$700 k–$2.0 M**
Foundry pilot production				
Development team	1 person	$150 k/year/person	0.5–1 year	$75 k–$150 k
MEMS fabrication	4 runs/year	$100 k-$200 k/run	0.5–1 year	$200 k–$800 k
CMOS fabrication	2 run/year	$50 k/run	0.5–1 year	$50 k–$100 k
Vendor services	3 vendors	$50 k/vendor	0.5–1 year	$75 k–$150 k
Subtotal (rounded up to nearest $100 k)				**$400 k–$1.2 M**
Total (rounded up to the nearest $1 M)			**2.5–5 years**	**$3 M–$7 M**

experience and charge a premium for their services. That expertise may be worth the expense because of the potential for reducing risk and overall project timelines.

We also assume that CMOS fabrication costs benefit from a standard foundry process, and that all the development stages leverage off-the-shelf packages. Table 6.1 does not include the cost of a custom CMOS design which can be more than US$1 M. We also exclude developing custom packaging, which can cost tens of thousands of dollars to hundreds of thousands of US dollars, depending on the type of package and its complexity.

The Duration Range column assumes that your technical and business development proceeds without any major interruptions, such as between fabrication runs and follow on vendor steps. For example, if you were to discover a material compatibility issue that requires sourcing a new material from vendor that still needs to be qualified, that adds time. If your funding source sets milestones with lengthy review processes as a requirement to release additional funds, that adds time.

The development cost model produces a range in total cost which is appropriate due to the ambiguities and uncertainties present at the beginning of a MEMS development effort. Although executives and accountants prefer to think of a budget in terms of one definite number, at the outset of a product development, it is more realistic and practical to present estimates in probabilistic terms, such as, "It's 90% likely that the development cost will be US$3 M–7 M." Alternatively, use the estimate to identify an upper bound on cost, such as, "It's 90% likely that the development cost will not exceed US$7 M."

Executives and accountants will no doubt chafe at this approach; however, intelligently assessing the viability of a company and its business model requires understanding and accounting for the uncertainty on a development cost estimate. A robust business model would enable a company to be profitable no matter whether the development cost turned out to be as low as US$3 M or as high as US$7 M. A business model that needs development cost to be capped at the lower end of that estimate, US$3 M, because of known market constraints that limit product unit price or unit volumes, will probably fail. In other words, to be successful, a business model should not count on everything going perfectly during development!

As development proceeds, however, model inputs and assumptions will become better defined, so the development cost model should be revisited frequently. By the time you finish the advanced prototypes stage, the high uncertainty associated with technology discovery and implementation will have been eliminated. What lies ahead will be execution with lower uncertainty. At that later stage, there will finally be enough information available to estimate remaining development costs with more certainty.

Reducing Development Timelines

After building a development cost model, you may identify some numbers that you wish to improve. Regarding development timeline, we frequently hear from our clients that their investor demands development of a MEMS product in under 2 years. The investor cannot wait 3–5 years for the product because of a time-sensitive window for the market opportunity. Assuming the technical requirements of your product can be met in a short period of time, your answer to an impatient investor could be yes. But you will certainly need to spend more money, and faster, to pull it off.

Additional development budget can be leveraged in several ways to shorten the timeline. The best leverage is to use it to parallelize all development tasks, which means that a larger team must be employed. Running development tasks and wafers concurrently carries risk of inefficiencies when information feedback loops are not quite closed. Going faster means that development money will be spent faster, and also that more of it will be consumed by the inevitable inefficiencies.

Table 6.2 outlines some different options for using money to accelerate development, the effect they can have on the development timeline, and estimated costs for the advanced prototypes stage. Chap. 15 provides additional strategies for balancing time, cost, and device performance in fab processing to further refine your development cost model.

The best practices and strategies outlined in this book are all aimed at reducing risk, and therefore, also cost and timeline. There are a few strategies which

Table 6.2 Methods for leveraging additional budget to reduce the MEMS development timeline in the advanced prototypes stage. Table based on 2020 Silicon Valley rates (US$); when applying this model, update numbers for the geographic location, device application, and level of complexity.

Timeline risk reduction method	Effect on timeline	Range of budget needed
More FEA modeling	Better understand your device and reduce the number of iterations in fab processing	<US$100 k
More fab short loops	Lower the risk in your full process flow and reduce the number of iterations in fab processing	US$50 k–100 k
Parallel design development and processing	More than one design is developed at the same time; best design wins and others are later abandoned	US$200 k–400 k
Parallel wafer starts; process more wafers	Increasing the number of wafers provides buffer or staging for when processing does not go as planned so that you do not have to start over from step one	US$100 k–200 k
Parallel work on packaging and electronics	Develop all system components in parallel	US$20 k–US$100 k
Vendor rush fees or other bonus payments	Save time at each process step to reduce the overall schedule	US$20 k–40 k
Have vendors build more tooling, add equipment, and increase capacity to enable parallel processing of your jobs	Processing can be easier and faster with more resources at the vendor	US$10 k–500 k
Recruiting engineers faster by paying finder fees and recruiters	Get the right team on board faster and then use that team to its fullest potential	US$20 k–200 k
Invest in on-site metrology and test equipment	Enable engineers to test silicon on site and get answers quickly to speed up cycles of learning	US$20 k–150 k
Pay for expertise and do not try to figure out everything on your own	Reduce the time to come up to speed on new techniques by bringing in consulting experts and integrating them with your staff	US$20 k–100 k

may reduce timeline and actually cost less. The following practices and respective chapters are highlighted for their ability to reduce both development timeline and overall cost.

- Licensing or buying intellectual property (IP) (Chap. 7).
- Using modeling to speed cycles of learning (Chap. 10).
- Using standard process modules or foundry process platforms (Chap. 11).
- Incorporating commercial off the shelf (COTS) components (Chap. 13).

Another way to reduce development timeline and cost, as well as to lower the final unit product cost, is to reduce the number of fabrication steps or operations your product requires. Your product and MEMS designers should be aware of this strategy when designing their initial proof-of-concept and advanced prototypes. Spending some time and money to explore options for more efficient manufacturing while in early development can save significant money over the long term.

Planning Staged Development and Go/No-Go Gates to Control Budget

The stages of development outlined in Chap. 3 exist, in part, because of how the MEMS industry is organized; where early-stage research and development can be done and how foundries must be managed for efficiency in volume production. There will always be some inefficiencies moving from one development stage to another when transitioning between the different partners and vendors. If executed well, however, these development stages can surface risks early before too much time and money is spent.

A key cost-saving strategy is to plan for design iterations early in the development, when the cost and impact of design changes is lower. This is one reason it is more cost-effective to use a development fab versus a foundry for building prototypes, as described in Chap. 17.

This strategy is illustrated by the MacLeamy curve, which was developed to show the cost of design changes in building design and construction, shown in Fig. 6.1.

Curve 4 illustrates an idealized design process where effort is concentrated earlier, when the ability to make changes is high and the cumulative cost of making those changes are low. Budgeting for this effort early in the process is critical to taking advantage of this relationship. Unfortunately, people are often unwilling to spend time and budget on concept studies and design explorations, instead being tempted to go straight to building in order to validate their ideas. If you take this approach, you may end up following Curve 3 and spending a lot more money. The MacLeamy curve suggests there is higher leverage in budget spent on early development stages because finding design flaws early prevents having to correct those errors during manufacturing, when making any change is very costly.

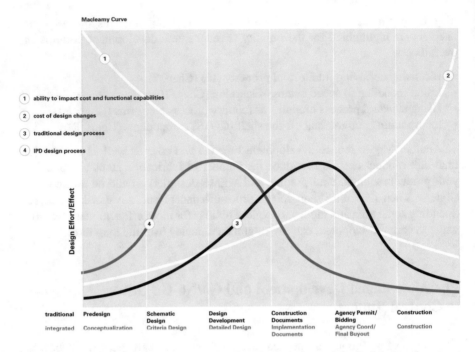

Fig. 6.1 MacLeamy curve: investment during the criteria design and detailed design phases prevents costly changes later in product development using Integrated Product Delivery (IPD, also referred to as the preferred design process) [1]. Reproduced with permission from the American Institute of Architects, 1735 New York Avenue, NW, Washington, DC, 20006

Staged development compels you to systematically explore your design and to validate your technology and approach before spending the next round of funds. Simply throwing a lot of resources and warm bodies at a design problem may work for software development but does not always help in MEMS. MEMS development is too multivariable, complex, and expensive to take this heavy-handed approach. A healthier way to "fail fast" in MEMS, no matter your company size, is to expend resources early in development to explore your design space and figure out the fatal process flaws or "cliffs" as soon as possible. In MEMS development, the aim is to identify and figure out your real development challenges early on. Define and set go/no-go gates for development stages to find fatal design flaws or major risks quickly.

In particular, dividing the advanced prototypes stage into specific, measurable milestones can help. If your product might have multiple technical challenges to overcome, you do not necessarily have to solve them all at once. Identify those which enable you to demonstrate the viability of the device and also those which could prove to be serious challenges. Starting with these challenges can be a smart approach if funding is limited—successful early results can yield additional funding. An added bonus to a staged approach, if you do it right, is that you can leverage the initial work to build foundational knowledge, and perhaps even to develop IP and trade secrets around the technology.

Go/No-Go Criteria for Moving to the Next Stage of Development

Breaking down development into bite-sized milestones provides a progress measuring stick. Take time to figure out and discuss your specific go/no-go criteria with colleagues. Success is not always clearly discernible in early stages; there may be nuggets of success amid the scrapped ideas. Development takes critical thinking to recognize which nuggets are strong enough foundations to support further development. This includes not falling in love with your early ideas; if data prove them weak, adapt as quickly as possible and move on to new ideas and more robust processes. If you are repeatedly failing to reach your development plan milestones, it is time to get out or go back to square one and reassess the design, business plans, and so on.

A good approach to choosing milestones is to first determine which criteria will be technical and/or business or market-based. Engineering teams typically focus on technical criteria as a measure of success. However, cost, profitability, customer demand, and so on are ultimately going to affect the success of a product and your business. Your go/no-go criteria should also include business criteria at all stages of the development plan. Table 6.3 lists example deliverables that can be used at go/no-go checkpoints to assess your progress.

Know your Company's Culture and Position to Inform your Development Plan

No matter whether in a start-up or a large established company, you need to demonstrate that you have reduced risk and demonstrated enough feasibility to justify further investment in the next stage. Taking a casual approach to development such as "swing for the fences" or "go big or go home" only works if there is lots of budget available and your company culture is very tolerant of failures.

Most company cultures are conservative, plan carefully, and need to see measured progress in order to continue to fund product development. Know which type of company culture you inhabit, or which type you want to create within your start-up. Design your development plan and go/no-go gates to meet the cultural expectations of your company.

The position your device has in the marketplace will also affect your development strategy. If there is market pull for your product, then the development timeline must be fast enough to meet the window of market demand. If you are pushing technology to the market, you need to adopt a long-haul mindset, and the development plan should include additional time to iteratively tweak the product so that it will eventually be accepted by skeptical customers (see Chap. 4).

In addition, if you know your customers will demand supply-chain redundancy, start planning to develop it as early as possible. If your company will be making MEMS components for use in complex products, especially those being

Table 6.3 Example deliverables for each development stage

Technical deliverables	Business deliverables	Risks eliminated
Proof of Concept		
Technical product requirements identified	Business product requirements identified	Product is well-defined for marketplace
Feasibility confirmed	Patent landscaping analysis	Patent infringement risk minimized
System design and simulation	Initial unit cost model validated	Performance specs can be met (verified by simulation)
Proof-of-concept prototype demonstrates some functions	Initial development plan cost model	Initial process integration complete
Initial package, electronics and software design	Initial analysis of long-term supply chain needs	Initial product integration complete
Initial plan for test		Initial test plan complete
Advanced prototypes		
Technical product requirements confirmed	Business product requirements confirmed	Product is viable for marketplace
Advanced system design and simulation models	Updated unit cost model	Predictive model accurately captures prototype device output
Advanced prototypes demonstrate nearly all functions	Updated development plan cost model	System specs are being met
Confirmed package, electronics and software design	Long-term supply chain identified	Product unit cost requirement can be met
Confirmed back-end processes and testing	Long-term supply chain identified	Product unit cost requirement can be met
Foundry feasibility		
Feasibility prototype is functioning properly; design and process frozen	Updated unit cost model	Design is manufacturable at foundry and meets performance specifications
System prototype is functioning properly; all design and manufacturing methods frozen	Long-term supply chain confirmed	Product unit cost requirement is on track
Initial samples available	Initial customers secured	Product is validated for marketplace
Reliability testing and quality control	Regulatory processes begin	Field failures and liability minimized
Foundry pilot production		
Wafer process window and cornering complete; CpK data	Finalized unit cost	Product unit cost requirement is met
Package, electronics, software complete and quality controlled	Product business model validated	Product pricing confirmed
Regulatory approvals received	Marketing and sales begin	No regulatory barriers to future sales
Initial product for sale in low volumes	Initial revenue	Business model validated
Foundry production		
Product for sale in high volumes	Long term supply agreements in place	Long term business is viable
Ongoing yield improvement and quality control	Supply chain management	Warranty return and product liability minimized

manufactured at high volume, you need to demonstrate security of supply to your future customers. Early in a product development, security of supply means having the foundry capacity to ramp up production volume to meet growing demand for your MEMS product. An OEM enjoying exploding popularity and demand for its product would never want to halt assembly lines to wait for one lagging MEMS component.

This means that during development, your company will need to understand and frequently reassess the long-range supply requirements for your MEMS component. If initial development occurs in a small fab using 150-mm wafers but you know your customers will eventually need the capacity of a high-volume, 200-mm-wafer foundry, your company must have a plan, schedule, and budget to eventually migrate manufacturing from the 150-mm foundry to the 200-mm foundry (see Chap. 17). Large OEM customers may be unwilling to buy your MEMS component until stable manufacturing has been achieved at a suitable high-volume facility, or perhaps even at two different facilities to enable security of supply by having a second source. Because of the long time and large investment needed to bring up a process at a new foundry, the development plan and budget must include paying for manufacturing at both the 150 and 200 mm facilities during a transition period that could last more than a year. Your company's cash flow and business model must account for this scenario.

With all these factors to consider, keep in mind that your personal or your team's definition of success may not be the same as that of your company's leadership. Work to align the technical and business goals, and you will increase your chance of success. For example, estimate development and production costs of different potential processes to provide options that may be more appealing to others on your team. In our practice, we have seen technically successful products never make it to market because they did not meet the profit margin expectations of the company's leadership.

Recouping Development Costs when Things Do Not Work out as Expected

The go/no-go gates of a well-structured development plan are designed to stop weak designs from draining a company's precious development dollars. If your development hits a no-go, it is not the end of the world. Product development is a very Darwinian process: only the strongest and best-executed ideas or technologies can make it out of development and into production; the rest will not. Investors know this, too.

So when you hit no-go, it is time to stop work and come up with a new plan. There may still be a way to restart your product development with a fresh idea, or it might be time to cut your losses and try to capture some value from your development to date. In Silicon Valley business jargon, to "pivot" means to abandon your original business model and/or product idea and head in a different direction, re-purposing the company's core technology in a different product embodiment, or for a different end use or market, and/or with a different business model. It is a healthy acknowledgement that even though you did not get to where you thought you would, the time and effort spent can be salvaged for a different purpose.

After a failed development, it might take a little time to figure out what that next idea or product could be. Use that time to document all the technology developed to date (see Chap. 16) and to file patent applications in order to preserve as much value of the development as possible. The technology might have to sit on the proverbial shelf until new ideas arrive or the market or company conditions improve. In the meantime, consider the following actions:

- Consider what new products may be created from the technology. A famous product that evolved from an adhesive development gone awry is 3 M's Post-It Notes [2].
- Retreat and regroup. Scale back your company to a size you can financially sustain while looking for a new path forward or additional funding.
- Repurpose the team and any facilities. If you built a great team of people, use them to develop something new. The right mix of people and vendor relationships can be easily leveraged.
- License or sell developed IP. It may be possible to pivot from a product to a service business model by offering your team's services to customize your IP to suit a customer's needs.

Summary

Building a development plan cost model can seem like a daunting task. There are a lot of variables and unknowns at the start of development. One way this can be tackled is by building a level-of-effort estimate where the number of people and projected process runs can be used to estimate development costs instead of by a task-based estimate built from the bottom up. Once you have a first model, you can start to make informed decisions on how to best spend your available funds to reduce risk and timelines. The model can also be used to evaluate expenses incurred as you approach milestones and to effectively communicate with and manage the expectations of executives and investors. It is important to remember that building a cost model is not a one-time effort. Throughout each development stage, you will need to refine the model based on your experience and increased knowledge of your MEMS product.

References

1. Institute of Architects (AIA) and the AIA California Council (2007) Integrated project delivery: a guide. American Institute of Architects, Washington, DC
2. 3M (2013) History Timeline: Post-it® Notes https://www.post-it.com/3M/en_US/post-it/contact-us/about-us/. Accessed 5 Aug 2020

Chapter 7
Leveraging Third-Party Intellectual Property to Accelerate Product Development

After reading Chap. 6, you might be sighing over the long duration and large budget needed to commercialize a MEMS product and wondering if any shortcuts might be possible. The answer, fortunately, is yes. Leveraging third-party intellectual property (IP), whether by license, purchase, or freely available, can leapfrog your development to a higher level of maturity.

There are several sources for third-party MEMS IP and technologies, each of varying maturity level and price. These are the most common sources, listed in order from less mature to more mature technologies:

- Academic institutions or universities.
- Research laboratories or institutes.
- Other commercial companies, including foundries.
- Public domain knowledge.
- Components for purchase or "white-label" technology.

This chapter describes these sources for MEMS third-party IP, the advantages and disadvantages of each source, and how best to deploy it in your product development efforts.

Academic Institutions or Universities

To successfully leverage academic research requires the same acumen as judging a diamond in the rough as well as the willingness, time, and resources to polish it into a valuable gem.

The technology available for license from universities will be very new and relatively untested, at Technology Readiness Levels 1–3. The focus of academic research, after all, is discovery and novelty, not manufacturing, repeatability, or reliability. Graduate student researchers focus on publishing papers and writing a

© The Author(s), under exclusive license to Springer Nature Switzerland AG 2021
A. M. Fitzgerald et al., *MEMS Product Development*, Microsystems
and Nanosystems, https://doi.org/10.1007/978-3-030-61709-7_7

thesis, not productization. For these reasons, academic research should never be regarded as a "turn-key" technology solution. Instead, it may provide your company with a head start on a novel technology that has the potential to create exciting new products. MEMS companies that have very successfully leveraged academic research to create commercial products include Chirp Microsystems (a subsidiary of TDK), Vesper, SiTime, and CardioMEMS (acquired by Abbott).

While developing their new technology, university researchers may have fabricated a handful of proof-of-concept prototype devices that produced exciting scientific results. Even if functional prototype devices do exist, understand that duplicating any early prototype, let alone achieving consistent device performance to a level suitable for commercialization, will still take significant investment and work. Most academic prototypes are made using manual or unorthodox fabrication methods, by whatever tools happened to be available, for the goal of demonstrating a new concept. Before it has a chance to become a product, the design and fabrication of the prototype must be translated to make it compatible with commercial, scalable manufacturing methods. This translation effort takes place in the advanced prototypes stage (see Chap. 3) and may take years and millions of dollars, particularly if the device design needs to be completely reengineered [1].

Universities are generally eager to license technology developed by their faculty and students in order to recoup research and facility costs. Most universities have an "Office of Technology Licensing" (OTL) or similar group, whose main purpose is to facilitate and grant licenses of patented academic research to commercial companies.

Due to the number and diversity of patents generated at a research university, effective cataloging and marketing of these assets by an OTL is difficult. You will never see advertisements for promising university technologies, nor would you be likely to stumble upon these technologies, unless you regularly attend academic conferences or have a personal network of academic colleagues. If you might be interested in licensing university technology, you will most likely need to go hunting for it.

One way to find academic technology for license is to comb through research journals and conference proceedings. This requires some time and effort; however, the papers you find will also reveal important information about the maturity of the technology. As you evaluate papers, look for detailed fabrication process descriptions, photos of operational proof-of-concept devices, and substantial data sets. When you find an interesting and potentially useful paper, simply contact the authors directly, or their institution's OTL, to begin business discussions.

The main advantage of licensing an academically developed, patent-protected technology is that it provides an opportunity to acquire truly novel technology at quite a low price. Many universities greet interest from commercial companies with enthusiasm and eagerness to reach a deal. License terms may be quite generous, such as no upfront fees and royalties due only upon successful sale of products incorporating the technology. Universities might offer some measure of exclusivity, usually by defining a specific market application(s) and/or a timeline for exclusivity. Rarely would they grant blanket exclusive rights to a technology.

You must conduct thorough due diligence to accurately judge the technology readiness level and to understand what is truly being offered. Evaluate all available documentation of the technology and make sure as much as possible is included in the licensing deal.

Academic papers and patents alone are insufficient documentation of a MEMS technology. You should request all computer files (mask layouts, CAD files, models, and so on), wafer process runsheets, process recipes, specification sheets from any special materials or components used, device assembly and test instructions, readout and test electronics schematics, and so on. (See Chap. 16 for more information on what items should be documented to smoothly transfer a technology from one party to another.)

The quantity and quality of documentation available may vary, according to the diligence of the professor and graduate student inventors, so be sure to understand what exactly you will receive from the university. If faced with inadequate documentation, it will be critical to secure consulting time from the professor, as well as key graduate students, if possible, to efficiently duplicate results and continue advancing the technology.

When licensing university technology, be sure to inquire about the option to receive some consulting support from the professor and/or to hire recently graduated students who did the hands-on work on the technology. If you are undertaking MEMS development for the first time, having these knowledgeable researchers available to assist with the transfer of licensed IP to your company will provide a significant business advantage.

Research Laboratories or Institutes

Government-funded or nonprofit research organizations developing MEMS and sensor technologies offer more mature technology than universities, usually at TRL 4–7. These organizations tend to have more advanced facilities and a dedicated, professional staff of engineers and scientists whose mission is to continuously develop and advance technology for license. The depth of knowledge, availability of documentation, and stability of fabrication processes will be superior to university research.

The Fraunhofer Institutes in Germany, CEA-Leti in France, Institute of Microelectronics in Singapore, the National Laboratories of the United States, and other similar government-funded research organizations aim to generate revenue from commercialization of their research and development work. Their respective governments also typically view successfully commercialized projects as ongoing justification for investing taxpayer funds in these institutions.

To best fulfill these objectives, they organize and develop their technologies into commercial-ready "platforms," often accompanied by other supporting components essential to a MEMS product, such as customized electronics and chip packaging. These organizations also have a dedicated business staff whose role is to market their technologies at tradeshows and commercial events, making their offerings much more visible to the commercial world than academic technologies.

To facilitate licensing of their technologies, most research organizations offer additional professional services, such as design customization, prototyping, and testing, for additional fees. They will usually be able to execute low-volume production of the platform technology within their own facilities. If your company will be licensing their technology for a niche market application, the challenging economics of fabricating low-volume specialty products (see Chap. 17) may strongly favor keeping your company's future production at the research institute.

If your company's MEMS product incorporates the institute's technology and will some day be sold in volumes higher than the institute's facilities can support (typically, more than one thousand wafers per year), you must plan to eventually transfer production out of the institute to a commercial foundry that can support the wafer volume, as well as any future growth. In this situation, during negotiation and prior to completing any license deal with the institute, you must investigate the true portability of their platform technology and processes from their facility to a third-party commercial foundry or fab facility. During negotiation, you must also secure the rights to transfer the institute's IP to future third parties. Bear in mind that even with full disclosure of all process and manufacturing details from the institute, transferring and stabilizing production of the technology at another commercial foundry or facility may still take more than a year of effort and over a million US dollars (see Chaps. 18 and 19).

When considering licensing technology from a foreign institution, you should also investigate whether any export restrictions might exist. For example, although the US Government laboratories, such as Sandia National Laboratory, have MEMS technology portfolios for license, they may be restricted from licensing to non-US–based companies.

As you should do when entering licensing negotiations with any organization, conduct thorough due diligence to evaluate the technology readiness level and to understand which items will actually be provided with the license. Depending on the institute, the license may or may not include rights to detailed manufacturing data such as mask files or process runsheets, or to peripheral technology essential to the product's function such as packaging, ASICs, and electronic circuits. Furthermore, access to and support from the institute's staff post-transfer usually requires additional funding beyond the technology licensing fees.

During license negotiation, it is also important to understand the ownership structure for existing patents or any new patents derived from the institute's original technology. Some institutes will demand joint ownership of any derivative IP, which your company's investors may object to because it would complicate your company's autonomy and valuation. Additional business contract terms such as exclusivity, warranty, and patent indemnification also need to be fully considered as part of valuing the licensing deal.

In particular, U.S. patents obtained as a result of research funded by the U.S. Government can be subject to additional restrictions.[1] The U.S. Government's

---

[1] Such patents are clearly marked with the language "This invention was made with government support under (contract number) awarded by (name of the federal agency). The government has certain rights in the invention."

rights will vary from contract to contract and may preclude manufacturing of the technology outside of the United States.

Although usually at a higher technology readiness level than academic research, do not assume that technology licensed from a research institute will be turnkey. If your company does not have its own expert team to receive and further develop the licensed technology, negotiate for ongoing support from the institute's staff.

Other Commercial Companies, Including Foundries

Many models abound for legitimately accessing technology belonging to another commercial entity. Some companies and foundries, usually large, well-established ones, may have MEMS technology available for license or purchase. The technology may be from their discontinued product, from an existing product that they hope to expand into new markets unrelated to their core business, or simply part of their strategy to win new manufacturing business.

For discontinued or obsolete product technology, a company may be willing to outright sell the technology (and all their old masks, tooling, and so on). Sometimes it may be possible to acquire technology as part of a bankruptcy proceeding or an asset-only company acquisition. In all cases, the technology will be in an as-is condition with no further support available from the original engineering team. To make good use and value of technology acquired in this manner, your company must have an expert team able to fill any gaps in knowledge or tooling.

Technology that is being used by its owner in an existing product may be licensable under specific constraints, such as over a limited period of time or in a restricted use (for example, in a specific, noncompeting market). In some cases, a company that owns a technology may be willing to modify or customize it for your new product's specific needs. Usually, this involves a nonrecurring engineering (NRE) fee in order to pay for the custom modifications, as well as a contractual commitment to purchase the newly customized part from them. Since the company's existing products and ongoing revenue would still depend on the technology, you can expect a high level of knowledge, documentation, and ongoing support of the technology.

A current example of this model is the Texas Instruments Digital Micromirror Device (DMD), which was initially commercialized by TI in the 1990s in the Digital Light Projection (DLP®) product. Sales of the DLP are still ongoing, and TI has since made its technology available to third parties. TI offers product-specific customization and manufactures chips for its licensees, using TI proprietary processes and its captive fab. TI also offers a family of DLP chipsets tailored for third-party applications such as machine vision, digital lithography, pico displays, and so on [2].

Some MEMS foundries also offer platform MEMS process technologies upon which a company may create its own new product design. Foundry process platforms offer a sequence of process steps and accompanying design rules, which are oriented towards creating a specific type of MEMS device, such as a motion sensor or a pressure sensor. (See Chap. 11 for more information on foundry process platforms.) The customer is solely responsible for creating the device design, so accessing foundry process platforms requires having expert MEMS designers. Recent examples of MEMS process platforms are the MIDIS inertial sensor process offered by Teledyne DALSA, a MEMS foundry in Canada; or the Sil-Via® Through Silicon Via offered by Silex Microsystems, a MEMS foundry in Sweden.

Licensing process IP from a foundry provides MEMS designers with the significant advantage of using a production-qualified, stable manufacturing process that is continually maintained by the foundry. In many cases, the foundry will also provide patent infringement indemnification for its process IP. Since much of the risk, timeline, and cost of developing a new MEMS device is derived from stabilizing a new fabrication process flow (see Chaps. 3 and 6), foundry platforms can offer a very enticing commercialization shortcut to expert MEMS designers.

Using the foundry's platform, however, requires that any product design incorporating the process technology be fabricated only at that foundry. Leveraging a foundry's platform therefore works best when your company does not expect to outgrow the foundry's wafer volume capacity and you are comfortable forming a long-term business partnership with that particular foundry.

Otherwise, having to switch foundries in the future could entail a complete redesign of your product in order to avoid transmitting any of the foundry's process IP and technology to one of their competitors. If you anticipate that your company's wafer volume might some day outgrow the foundry's capacity, you must address this potential issue during initial discussions with the foundry. In some cases, a foundry may allow a product to be "second sourced" at another foundry for additional licensing fees (See Chap. 18 for more information).

Public Domain Knowledge

Public domain knowledge, in the form of textbooks, publications, expired patents, or abandoned patents (which fall into the public domain when the assignees fail to pay maintenance fees), can be a rich source of technical information. Significantly, crafting a MEMS design to follow known public domain information is a business strategy that minimizes the cost of developing a custom, but not innovative, MEMS design. In some cases, it may also provide a shield against patent infringement claims.

Most novel MEMS devices are first developed by universities or government-funded research organizations who publish and disseminate their research to the public. A lot of useful information about a device or process technology can therefore be found online, in the academic literature, and in open discussion at conferences.

This public information may or may not be free to use in your commercial product. University professors, for example, often file patent applications before publicly disclosing their research at conferences or in journal publications, especially when obvious commercial applications exist.

Before implementing any information about a technology found in an academic paper or other publication, be sure to search both the issued patent and the published application databases of the US Patent and Trademark Office (USPTO) and the Patentscope database of the World Intellectual Property Organization (WIPO) for the author's name and organization, in order to identify any issued or pending patents.[2]

Consult with your company's IP attorney. They may need to conduct a "freedom to operate" analysis to determine if the paper's information is truly public domain information or may be protected by patents. Furthermore, the protection afforded by a patent is limited to what is written in the claims that issue, not by what is disclosed in the body of the patent, or in the claims written in a pending patent application. The protected claim scope for an issued patent is best interpreted by a trained IP attorney [3].

The long duration for any new technology to evolve from first invention to commercial product, and the difficulty of developing MEMS in general, means that sometimes MEMS patents expire just as the marketplace is finding good uses for that particular technology. When this is the case, there is an opportunity to leverage expired patents and existing public domain information in order to deliver your company's own customized version to awaiting customers.

An example of this situation is MEMS micromirror technology, which many commercial entities developed in the late 1990s in anticipation of demand for fiber-optic networking equipment. The hoped-for market demand never fully materialized at the time, and many companies developing micromirror technology went out of business, leaving their technology in the public domain [4]. Fast forward nearly 20 years later: long-awaited demand for micromirrors to be used in fiber-optic networking equipment finally occurred [5]. In addition, new interest arose in micromirrors for application in virtual/augmented reality (VR/AR) displays and LiDAR systems. Expired or abandoned 1990s-era patents could be leveraged to quickly customize micromirror components for these emerging market opportunities.

MEMS pressure sensor technology also has been especially long-lived and has a compendium of public information available for free use. Many MEMS pressure sensors sold today incorporate methods and designs patented in the early 1980s, which fell into the public domain long ago. Having your company's IP attorney conduct a freedom to operate analysis before deploying expired or abandoned patent information is still advised.

A major consequence of leveraging public domain information in your product's design is that it could impair ability to patent. An IP attorney should be consulted to identify what options, if any, may exist for patenting your product design based on public domain information.

---

[2] N.B. Only applications filed more than 12 months ago are included in these databases. In addition, an inventor can request that the USPTO not publish a patent application until it issues, in return for agreeing to file only in the US. As a result, it may be possible that patent protection has already been sought for a technology even though nothing appears in these databases.

From a practical business standpoint, however, inability to patent a device may be insignificant for some applications and perhaps quite a beneficial trade-off for the time and budget savings gained from leveraging public domain information. There are many ways to protect a business's competitive advantage besides patents, some of which are discussed below.

Components for Purchase or "White-Label" Technology

Being unable to patent may indeed be of little to no consequence in the case where the MEMS device is a component incorporated within a product having a system function derived from the interaction of multiple components. The novel and non-obvious combination of existing technologies may be patentable as a system design, even if the individual components exist in the public domain. Furthermore, a novel method of using a known device in an unexpected and nonobvious way might also be patentable (e.g., a method for performing brain surgery using a hammer).[3]

In addition, proper function of a MEMS device in the product application's environment may require special engineering, such as mounting, packaging, drive electronics, hermetic sealing, biocompatible coating, and so on. Any or each of these application-specific enhancements may be separately patentable, or they may be jointly patentable as a novel sensor system. In some situations, it may be possible to add significant value and competitive advantage to an existing MEMS device by applying special calibration techniques, unique packaging, and customized electronics and/or software (see Chaps. 9, 12, and 13).

This strategy of special proprietary enhancements to commercially available components has been successfully employed by several companies who sell MEMS microphone-enabled products, yet do not possess or make their own MEMS microphone chip designs. Instead, these companies purchase a "white label" or "blank label" MEMS microphone chip from the German company Infineon, and then add value and differentiation through proprietary acoustic packaging, codec electronics, and calibration. For example, microphone products sold by both AAC and Goertek incorporate Infineon MEMS microphone chips [6].

So while a MEMS device design itself may not be patentable, or perhaps was even purchased from a third party without any exclusivity, the unique combination of the chip with novel integration features can make a compelling and competitive product. Purchasing a generic MEMS component, instead of trying to design a custom one from scratch, can save years and millions of US dollars in product development effort. System-level patents may be sufficient to protect the product in the commercial marketplace.

---

[3] Witty example provided by James A. Walker.

Summary

Caveat emptor, the ancient Roman advice meaning "buyer beware," readily applies to modern-day technology licensing and purchase deals. Leveraging third-party technology instead of inventing your own can save significant time and money in the long haul of new product development. Selecting third-party technology requires clear-eyed and thorough due diligence in order to correctly assess its maturity and its usefulness to accelerating your product development efforts. If not in a ready-to-use state, be sure to also assess any investment needed for further development and to negotiate for ongoing support from the source's experts. More mature technology always commands proportionately higher prices due to lower associated risk.

References

1. Fitzgerald AM, Jackson KM, Chung CC, White CD (2018) Translational engineering: best practices in developing MEMS for volume manufacturing. Sensors and Materials 30(4):779–789
2. http://www.ti.com/dlp-chip/getting-started.html
3. Rockman HB (2004) Intellectual property law for engineers and scientists. IEEE Press, Piscataway, NJ
4. Ramani CM. Optical MEMS: boom, bust and beyond. *2006 Optical Fiber Communication Conference and the National Fiber Optic Engineers Conference*, Anaheim, CA, 2006, pp. 11 doi: https://doi.org/10.1109/OFC.2006.215864
5. Chollet F, Liu HB, Ashraf M et al (2009) Of light, of MEMS: optical MEMS in telecommunications and beyond. Sadhana 34:599
6. Clarke P (2020) Infineon is moving the growing microphone market. *EE News Analog*, March 30. https://www.eenewsanalog.com/news/infineon-moving-growing-microphone-market

Chapter 8
Organization Planning for Successful Development

We tend to imagine exciting new products as the work of a sole genius, a belief often perpetuated by enthusiastic feature articles and movies. While there indeed may be one person who sparks the creation of a new product or champions it within a company, bringing a new technology product to market requires a well-functioning organization of many, many people.

These days, technical products are much too complex for one individual to possess all the knowledge and skills required to realize the product, let alone to operate the business of making and selling the product. As product development proceeds, an evolving set of skills and knowledge become needed to keep pushing the technology towards maturity.

In this chapter, we describe the functional roles and skills essential at various stages of development. The exact partitioning and organization of these roles into individual job descriptions, and the leadership hierarchy for the team, will vary according to a company's existing structure or cultural preferences. Similarly, the number of people needed to fulfill each functional role will depend on several factors, including the complexity of the product being developed, the complexity of its end use application, and so on, down to such specific details as the labor laws in a business's particular municipality that might dictate a maximum number of working hours in a week.

We intentionally do not describe the roles or organization required to operate a business selling a mature product because that topic is already well addressed by books on generic product business organization. Instead, we focus on the organizational needs specific to MEMS development from proof-of-concept to foundry production.

The functional roles we describe below represent *a minimum set* that must exist for successful execution. We provide each set in the context of different company profile examples, each of which may be starting a MEMS product development at a different stage.

A. M. Fitzgerald et al., *MEMS Product Development*, Microsystems
and Nanosystems, https://doi.org/10.1007/978-3-030-61709-7_8

Proof-of-Concept Stage: Start-Up Company or R&D Group within an Established Company

For a team trying to develop a new type of MEMS component, there are two main roles to fulfill: MEMS engineering and business. In this stage, a new technology has been created, a product opportunity has been identified, and a business is beginning to coalesce around the product opportunity. This is the point at which a start-up company would be incorporated, or a business unit team might be formed within an existing company.

Figure 8.1 illustrates the minimum functions needed during early-stage product development. At the very beginning, and for a short while, a single person may be able to fulfill all of these roles. Once fundraising (or corporate budgeting) begins, however, there must be a bifurcation in roles to manage the growing workload. Another important business task is to engage with future customers to inform product definition (see Chap. 4). Preparing for and meeting with potential investors, corporate executives, as well as potential intermediary integrator or OEM customers requires dedicated resources. At least two people will be needed to keep both the engineering and business efforts moving, respectively.

With the arrival of funding (or allocation of corporate R&D budget), each of the two teams—MEMS engineering and business—need to further expand (Fig. 8.1).

The MEMS engineering team will expand faster initially, in order to add expertise and bandwidth in mechanical, electrical, and process integration design, as well as any proof-of-concept prototype fabrication. Even if prototype fabrication were to be entirely outsourced, someone on the engineering team would need to be a liaison providing input and feedback to the vendors.

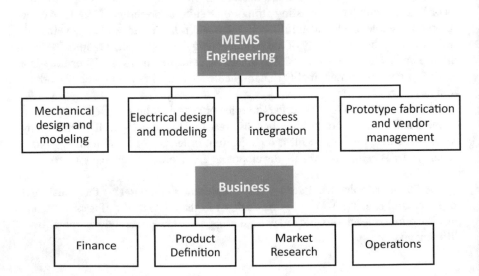

Fig. 8.1 Minimum functional roles required during the proof-of-concept stage

In parallel, the business roles will expand to include market research and product definition. In a start-up, the day-to-day finance and operations roles (accounting, legal, human resources, facility, and so on) must also be added to keep the company operational.

At this point, depending on the complexity of the MEMS component under development, the entire team may consist of at least four people. The main goals of the engineering team during this stage are to develop the technology and stabilize the function of the device. The main goals of the business team during this stage are to secure financing and validate the product and market opportunity.

Advanced Prototypes Stage: Start-Up Company or an Established Company Licensing MEMS Technology

Once a proof-of-concept prototype demonstrates adequate function on the benchtop (that is, attached to power supplies, external circuit boards, drive electronics, and so on), the advanced prototypes stage of development begins. At this point, a start-up company might be receiving its Series A funding from venture capitalists. To bring a new example company profile into the discussion, an established company might have licensed a proof-of-concept prototype from a university or research entity and is therefore beginning product development from this point.

The team must now focus on developing advanced MEMS prototypes and demonstrating how they might function within the complete product, such as a medical device or an automobile. This is the stage at which development work must include product or system integration.

The engineering team will continue to develop the MEMS product; its interface to its own package and electronics, forming a module; and how that module will interface to the end use product or application. Software development may also be needed at this stage.

Additional expertise, in the form of a product (or system) integration team, if not already present, must be added. The composition of the overall team (Fig. 8.2) will be determined by the end use market(s); for example, if the MEMS product is destined for automotive applications, the team requires expertise on the function, integration, and reliability needs of automobiles. If a MEMS product will be sold for use in invasive surgical equipment, the team must have expertise on biocompatibility, sterilization, reliability, and regulatory needs of medical devices. No matter what the end use application may be, the team also needs someone to focus on product engineering, meaning, continually evaluating the product from potential customers' points of view and making sure that it will meet all product requirements [1].

End use and integration expertise must be present on the team before developing any advanced prototypes. Product and system integration requirements must be defined and then propagated down to the die-level MEMS mechanical and

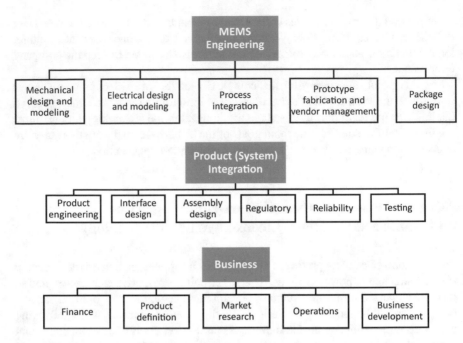

Fig. 8.2 Minimum functional roles required at the advanced prototypes stage in product development

engineering specifications. These integration requirements will inform items as detailed as choice of interconnect metals, size and location of bond pads, passivation materials, and so on.

Not having product and system integration expertise at this point in the development means that the MEMS product- and die-level designs will continue to evolve and harden in isolation. When external integration factors are finally considered several months or years later, there could be a serious mismatch between the die's existing design and what the product actually needs to be in order to be useful for a specific end use application (see examples in Table 8.1). Rectifying the mismatch would require costly and time-consuming redesign; the worst-case scenario would be that you have spent time and money developing a product that the end user cannot implement or will not want.

During this stage, all teams will need members with additional skills. On the technical side, more expertise in package and assembly will be needed as the product or system team begins to define the interfaces between the die and the product, and deeper expertise will be needed as development inevitably uncovers subtle physics and process interactions.

On the business side, a business development specialist will be needed to begin finding and interacting with potential customers and end users and laying groundwork for future sales. In particular, the business development and product integration leads need to continue talking with potential customers to elicit the respective

Table 8.1 A few examples of the risks and consequences of failing to consider system integration requirements at the advanced prototypes stage of MEMS development (Source: AMFitzgerald)

Item	Risk	Severity and consequence	Impact to development[a]
Bondpads	Bondpads in wrong location(s) for package interface or integration	Low; redesign one mask level	Delay: 1–2 months Cost: $1–10 K
Die layout, size	Die layout does not include correct features or dimensions for product integration	Low; redesign of entire mask set	Delay: 1–2 months Cost: $10–100 K
Biocompatibility or sterilization compatibility	Incompatible material(s) must be removed from MEMS die or encapsulated	Medium; redesign of process flow	Delay: 2–4 months Cost: $50–300 K
Operating voltage	System electronics do not supply correct voltage, so MEMS device will not function as well as intended or at all	Medium; redesign of electronics and/or MEMS device for compatible voltage operation	Delay: 2–4 months Cost: $50–300 K
Force or displacement of an actuator	MEMS actuator does not provide enough force or displacement for intended application	High; entire device architecture may need to be changed to accommodate different transduction methods; complete redesign of both die and process flow	Delay: >6 months Cost: >$500 K

[a]All costs in US dollars, based on typical costs in Silicon Valley at the time of writing

business and technical characteristics needed to make the product sellable and successful.

The entire team may grow to between 10 and 50 people during this stage, depending on the complexity of the product.

Foundry Development Stages: Start-up or Established Company

Once advanced prototypes have been stabilized and demonstrate the form, fit, and function desired by potential customers, then the final product development and integration begins. At this point, a MEMS component would be going through the stages of development at a production foundry and units would soon be available in larger quantities for reliability testing, customer samples, and so on.

The engineering team at this stage must include experts on product design (and is now large enough that it needs its own figure, Fig. 8.3). Mechanical and electrical integration, and package design efforts, will become focused on achieving

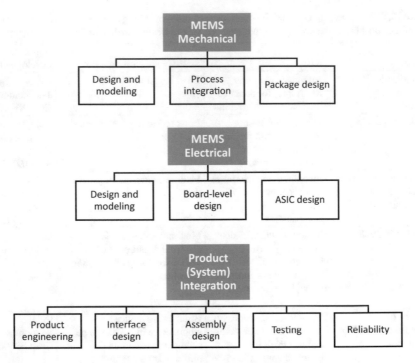

Fig. 8.3 Minimum technical functional roles required at product integration stage

customers' particular specifications. Depending on the product and its end use, a software team may also be needed. Testing and reliability evaluations that were previously focused on understanding the MEMS device will now evolve into product qualification tests, verifying that the product will indeed meet customers' requirements. In many cases, customers' requirements will also be dictated by industry test standards, or national or international regulatory requirements.

On the operations and business side (Fig. 8.4), a group devoted to manufacturing must now be formed, regardless of whether manufacturing occurs in-house or is outsourced (more on that in a moment). The manufacturing group should include expertise on all manufacturing steps, supply chain management, and quality control. The manufacturing group's focus should be on stabilization of existing methods, continually improving yield of manufacturing, and minimizing product recurring costs.

The manufacturing group should now become isolated from any parallel innovation efforts. Innovation requires constant tinkering and experimentation, which naturally antagonizes the stability and repeatability needed for efficient manufacturing. Ideally, the innovation and manufacturing teams should work in separate facilities. If that is not possible, then a carefully planned facility usage schedule and operating rules should enable the two teams to share the space without conflict.

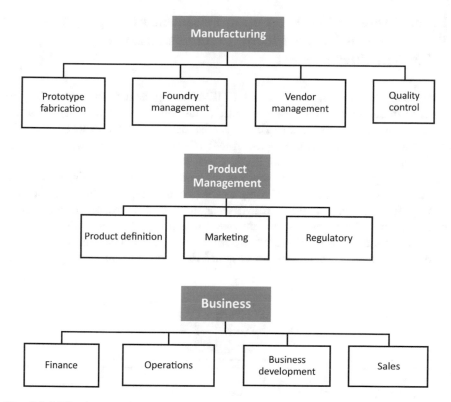

Fig. 8.4 Minimum operational and business management roles required at product integration stage

Additionally, this is the point at which a product management team should be formed. Their purview will be to manage current product specifications with respect to customer and market needs; to identify new product opportunities that may be achieved with the device technology; to create effective strategies for marketing the product; and, finally, to stay informed of and to manage any regulatory requirements and reporting.

Product samples may already be available at this stage, and with volume manufacturing imminent, sales are on the horizon. A sales organization is now needed. Depending on the end use market, sales people with direct experience and human networks in the end use applications, and a specific set of language skills, may be required.

During this stage, the number of people in each role will naturally increase as product sales grow successfully. The organization template may be cloned and replicated into new divisions or business units, as new generations of products become developed and then commercialized.

Product Integration: Start-up or Established Company That Purchases Third-Party MEMS Components for Its MEMS-Enabled Product

A company that purchases a packaged MEMS component from a third party in order to integrate it into a MEMS-enabled product will need only a subset of the functions and roles compared to a company that designs and builds its own MEMS component. In this scenario, since the MEMS is a purchased component, the design and manufacturing teams are not needed, and the technical team will be focused only on integration of the component with the company's product (Fig. 8.5).

As discussed in Chap. 7, purchasing a MEMS component is a very cost-effective strategy for acquiring and using MEMS technology. It also provides the major benefit of keeping the development team small, which would dramatically lower the amount of funding needed to launch a MEMS-enabled product. This strategy may also be used to bootstrap towards a custom MEMS component development. The

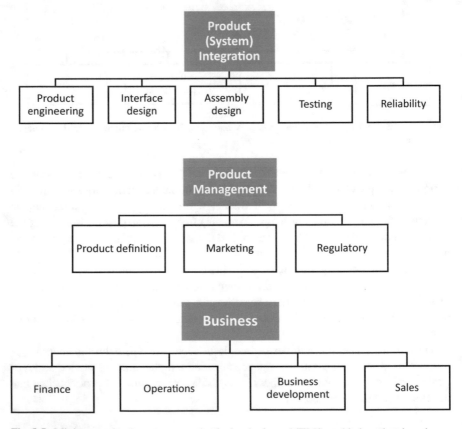

Fig. 8.5 Minimum roles for a company that is developing a MEMS-enabled product based on a purchased MEMS component

first-generation product would use a purchased MEMS component to minimize investment in development, to prove market viability, to sharpen product definition, and, importantly, to generate revenue. All of the knowledge and profit from the first-generation product could then be fed back into developing a second-generation product having an intelligently optimized, custom MEMS die.

External Team Members

In wafer and electronics manufacturing, due to costly capital equipment, need for special facilities, and need for personnel with unusual skills, outsourcing is a widely used development and operational strategy. Even large, vertically integrated companies use outsourcing to some degree. A major consideration for any MEMS company is to determine which portions of the development could be or should be outsourced and then to build up a supply chain. Hiring talented employees is a critical success factor for any company, but so is selecting capable and reliable vendors to execute key operations.

Outsourcing to Vendors

How and where to partition work between in-house (insource) and outsource teams requires thoughtful analysis. Before any trade-off analysis begins, perform a frank assessment of your company's IP, team, and resources. In particular, assess where your company is really adding value to the product with cost and time efficiency and low risk; any area where it is not should be considered for outsourcing. Understanding what knowledge or skills are your company's or product's "crown jewels,"—those which are truly novel and worthy of investment and protection—will help to guide decisions about what should be kept in-house.

Table 8.2 identifies common operations or services in MEMS product development and some reasons to consider outsourcing versus insourcing them.

Outsourcing is not an all or nothing proposition. Even within a specific manufacturing operation, it is possible to implement a combination of outsourcing and insourcing. One example would be fabrication of a chemical sensor; the MEMS wafer fabrication process could begin at a foundry and proceed to a certain point, after which the wafers would be sent to the sensor company, which then applies a proprietary functional coating in order to complete the device fabrication.

A common strategic error of young companies is to try to do and control too much themselves. Often this occurs because of the following factors: lack of understanding about which parts of the MEMS development are truly novel and special, leading to unwarranted anxiety about IP theft and therefore an aversion to outsourcing; lack of awareness of what capabilities are available in the wider vendor network; unfamiliarity with how to find and engage vendors; and sometimes, even a

Table 8.2 Examples of common MEMS product development operations or services and reasons to outsource or not. (Source: AMFitzgerald)

Operation or service	Reasons to outsource	Reasons to insource
ASIC design	Niche skill; many capable teams available having existing semiconductor foundry relationships	MEMS performance depends significantly on ASIC performance, requiring tightly coupled development (e.g., gyroscope)
MEMS design	The MEMS die provides a single function and is one of many components in your company's complex system product	The MEMS die is your company's product or completely enables the product
Packaging and assembly	Specialized robotic equipment required for cost-effective service; many capable vendors in low-cost regions	Low-volume production and/or the package contributes significantly to the performance of the product (e.g., implantable medical device)
Printed circuit board manufacturing and assembly	Many skilled, low-cost vendors available with automated equipment	Rare circumstances in which some aspect of the PCB is adding significant value to the product
Wafer dicing	Many skilled, low-cost vendors available with automated equipment	Rare circumstances in which some aspect of the dicing step is adding significant value to the product
MEMS wafer fabrication	Many foundries available with a broad range of capabilities; very capital intensive	Low-volume devices, especially those which are also low-process complexity (e.g., specialized microfluidics); when unusual processes are required that are not available at foundries; protection of trade secret processes
Wafer-level electrical test	Many OSAT vendors available with automated, high-throughput equipment in low labor cost regions	Fast turn time needed; need for customized equipment to apply physical stimuli; unusual or delicate testing conditions required
Functional testing of product	OSAT vendors available with essential equipment and special skills in low labor cost region	Specific functional test needs or sensitivity of data
Reliability testing	Vendor with essential equipment and special skills	Due to specific functional test needs and sensitivity of data, this is often done in-house

desire to control certain aspects of the product because they are intellectually fascinating. A wise entrepreneur knows not to spend time reinventing the wheel; in the long run, it is always cheaper and faster to hire skilled vendors to get the job done properly the first time rather than to try figuring it out by yourself.

Ultimately, the main purpose and benefit of outsourcing is to provide strategic leverage in at least one of the three key factors underpinning all business operations: cost, time, and risk.

Vendor and Supply Chain Management

A MEMS development effort needs a steady stream of specialty supplies and materials to keep proceeding, and it can also significantly benefit from leveraging external expertise and capabilities. Locating and qualifying vendors and then managing those business relationships is an important role on a development team. In the earlier stages of development, the vendor management role may be filled by technical staff who also have the detailed expertise to interact and work closely with vendors. Over time, especially once the production stage is reached, the role should be filled by operations and supply chain experts. Ideally, these people should have deep knowledge of the global industry ecosystem and experience with outsourcing business practices [2].

The MEMS and semiconductor industries also occasionally experience disruptions due to factors beyond the control of any company, such as shortages of key raw materials due to conflict or new regulation, price spikes or long lead times due to unusual demand, or supply disruption due to natural disasters like earthquakes and more recently, pandemic. If your company will be supplying high volumes of product to an intermediary integrator or OEM, a supply chain management team will be required. Experienced supply chain management personnel will be quick to comprehend the ramifications of these disruptions and adept at finding alternate sources or work-arounds to maintain the necessary flow of materials and services to keep your production moving.

Summary

As product development progresses from the proof-of-concept stage all the way to production, your company's essential team roles will shift and the roster of employees will expand. Identifying and hiring talented employees for key roles, especially in the arcane field of MEMS, takes time and money. Having foresight on which roles will be needed as development progresses enables strategic planning for recruiting and hiring.

Even though they are independent entities, external vendors and service providers are vital members of a MEMS development team. They bring expertise, equipment and operational capacity. A company should thoughtfully build its development team to strategically utilize the capabilities of both internal and external members.

References

1. Cagan M (2017) Inspired: how to create tech products customers love, 2nd edn. John Wiley & Sons, Hoboken, NJ
2. Hugos MH (2018) Essentials of supply chain management, 4th edn. John Wiley & Sons, Hoboken, NJ

Part III
Technical Requirements for a Viable Product

Chapter 9
The MEMS Product: Functional Partitioning and Integration

While the MEMS device is the distinguishing and strategic component in a MEMS product, it is only one part in an overall system. The MEMS device cannot function properly without its accompanying components. The "S" in the acronym MEMS stands for systems, and in many parts of the world, the preferred word for these products is "Microsystems."

A MEMS Product Is a System

The minimum MEMS product includes a MEMS device, package, and drive/readout electronics (see Fig. 9.1). Software is also an useful and valuable component and is often included. Wireless MEMS products, such as for Internet of Things (IoT) applications and biomedical implants, may also have a battery power subsystem and a wireless communications subsystem. Combination sensor products may include multiple MEMS devices and their associated packages and electronics. One example of a "combo" sensor is an Inertial Measurement Unit, which may be composed of a MEMS gyroscope, MEMS accelerometer, pressure sensor, and magnetometer.

Each of the components in a MEMS product provides significant functionality and therefore value to the overall system. Table 9.1 gives a summary of each component's functions. Understanding the role of each component in the system enables the development team to optimally partition functional roles among the components, which can lower the overall cost, time, and risk of the MEMS development and its productization. MEMS engineers should specify each of these components at the beginning of a development program and develop all of them in parallel. Focusing development solely on the MEMS device while neglecting or postponing work on the other components will significantly delay and increase the costs of the MEMS product's commercialization (see also Table 8.1). Chapter 13 discusses parallel co-development of these three parts of the microsystem in further detail.

© The Author(s), under exclusive license to Springer Nature Switzerland AG 2021   95
A. M. Fitzgerald et al., *MEMS Product Development*, Microsystems
and Nanosystems, https://doi.org/10.1007/978-3-030-61709-7_9

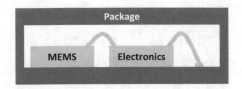

Fig. 9.1 A conceptual illustration of a cross section of a package-level integrated MEMS product, composed of the MEMS chip, electronics chip (ASIC), and the package

Table 9.1 Functions that are typically partitioned to each component in a MEMS product

Microsystem component functions		
Package	Electronics	Software
Environmental protection	Power	Feature extraction
Electrical protection	Readout	Event identification
Mechanical protection	Signal processing	Datalogging
Access to sensed/actuated phenomenon	Calibration	Rapid product fixes
Electrical connection to system	Compensation	
Mechanical connection to system	Wireless communications	
Labelling	Energy storage management	
Handling		

In this chapter, we review the three main system components present in MEMS products:

- Package.
- Electronics.
- Software.

Passive (unpowered) MEMS products, such as microstructured surfaces, optical gratings, and some microfluidic chips, may not need electronics or software, but may require special packages to enable or enhance function.

Package

The term "package," when used by MEMS or semiconductor engineers, refers to a structure that encloses a silicon chip. The MEMS package has several important functions. The package is the mechanical and electrical interface between the MEMS chip and the greater system. It is also the interface between the MEMS chip and the sensed/actuated phenomenon. Finally, the package protects the MEMS from the operating environment (Fig. 9.2).

The package has a difficult role in that it must reject environmental hazards, while selectively allowing electrical signals and the phenomenon with which the MEMS chip interacts to pass from outside the package to the MEMS chip.

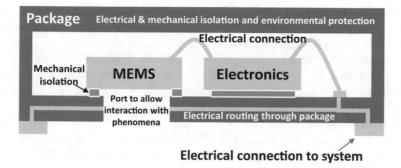

Fig. 9.2 Cross-section illustration of some of the functions provided by a MEMS package

The package must meet the requirements of the MEMS chip as well as the application environment. For example, a package for a pressure sensor has different requirements from a package for an accelerometer; the former requires a hole, whereas the latter does not. And likewise, a package for a MEMS to be used in a smartphone has different requirements from a package for a biomedical implant. The combined requirements of a particular MEMS chip with a particular application often means the package is not available off-the-shelf and must be custom developed.

The package plays many roles, and the package can also add significant value to the overall MEMS product. As a result, the time and resources allocated to the package's development should be viewed not as a cost, but as an investment.

Environmental Protection

Bare MEMS chips are delicate due to their microscale features. They are vulnerable to many hazards in the surrounding environment, including humidity, water, electrostatic discharge, rough handling, dust particles, and much more.

Dust particles and humidity are common environmental hazards. The typical feature in a MEMS device is on the micrometer scale. Specks of dust are often 10 or 100 times larger. A speck of dust at the wrong location can cause a MEMS device to fail. Likewise, a droplet of water can be much larger than the features inside a MEMS chip. Water may cause the MEMS device to electrically short or to corrode, and water can cause stiction between moving parts in the MEMS device [1].

By enabling the MEMS to operate in its environment, the package provides a critical functionality. Understanding that it is the package's role to enable the MEMS to operate in its application environment allows the development group to partition that role to the packaging. For example, consider a MEMS sensor that needs to function in a corrosive environment. While it may be possible to design a MEMS device that can withstand exposure to a corrosive chemical, it is usually more cost effective to delegate chemical resistance to the package. One reason is greater flexibility in the material selection for the package. The package material can be chosen from materials that are inert to the particular chemistry of the application's environment, such as glass or Teflon.

Another example is an implantable medical device. There are many challenges to operating a delicate, tiny electromechanical device within the human body. For example, the microsystem must be protected from the corrosive, saline environment of the human body. It must also avoid or tolerate invasive tissue growth and immune system attack. Furthermore, the product must be able to withstand sterilization methods and handling during implant. All of these environmental issues are usually most cost-effectively solved by the package, which then allows the MEMS device to be focused on its strengths, sensing and actuation.

Electrical Connection

Another basic function of the package is to enable the MEMS device to electrically connect to other components within the microsystem as well as to others outside it in the greater product. This means that the package must be selective about what it allows in and out; for example, it must keep out phenomena like dust and water but permit electrical signals to pass through. Allowing the passage of electrical signals can compromise a package's function to protect the MEMS device. Electrical signals are passed through the package by conductors, referred to as the "electrical feedthroughs." The electrical feedthrough is often a point at which package protection becomes compromised. For example, water has a notorious ability to penetrate the smallest spaces, so a MEMS device that will operate in a humid or aqueous environment requires a package with engineered hermetic electrical feedthroughs.

Interaction with Phenomena

MEMS devices are usually either sensors or actuators, which means they must interact with the phenomena that they are sensing or actuating, typically through ports in the package. Again, this requires the package to selectively allow the sensed or actuated phenomena, while rejecting environmental hazards like water or dust.

Access to some phenomena is straightforward, such as with inertial sensors. Since inertia cannot be "shielded," inertial sensors may be completely encapsulated in a package without any open ports and still able to sense inertial changes.

Access to other phenomena can be challenging, such as for microphones. Sound is a relatively weak phenomenon. The energy of a sound wave from normal conversation for 1 s is 10^{-6} J [2]. In comparison, a paper clip (1 g) falling 1 meter has an energy of 10^{-2} J, and sunlight falling on a 1 m^2 area for 1 second has an energy on the order of 1000 J [3]. The package must allow sound to reach and interact with the

MEMS microphone's membrane, while still protecting the membrane's surface. In this case, an opening in the package allows sounds to travel from the environment to the microphone. The opening, however, leaves the MEMS die vulnerable to water vapor. Protection from vapor is instead partitioned to the MEMS die; that is, the MEMS microphone chip is specially coated so that it tolerates humidity.

Electrical Isolation

MEMS sensors often have tiny signals that are in the microvolt, picoamp, and/or femtofarad range, or smaller. These signals are easily overwhelmed by spurious electrical noise in the external environment. Automotive MEMS devices, for example, may need to operate near the engine, which is a source of electromagnetic noise. The package can provide electromagnetic shielding from noise sources by surrounding the MEMS and ASIC with conductive layers or meshes to create a Faraday cage. The Faraday cage protects the delicate analog sensor signal until the ASIC converts it into a more robust digital form (the ASIC is discussed more in Chap. 13).

Mechanical Isolation or Damping

MEMS devices are mechanical in nature, so they also need isolation from spurious or harmful mechanical inputs. Vibration and shock from the end use application's environment, as well as during handling and transport, can cause MEMS structures to vibrate, collide, fracture, and/or generate particles. The package can provide a degree of mechanical isolation by damping external shocks and preventing damage to the MEMS die. The package could also contain an inert pressurized gas or a liquid that provides viscous damping against spurious mechanical motion. This may also be employed to engineer the dynamic response of MEMS devices, such as to prevent excessive ringing.

All MEMS devices have sensitivity to temperature due to material coefficient of thermal expansion (CTE) as well as mismatches in CTE to attached materials. For example, in most systems, the MEMS is mounted in a package that is mounted to a printed circuit board (PCB), which can have a CTE that is 6 times greater or more than a silicon-based MEMS [4]. As the PCB undergoes heating and cooling, the PCB exerts mechanical forces on the package, and in turn, the MEMS die. These forces can cause distortions in the MEMS output signal. The package, if designed appropriately, could serve as a mechanical buffer between the MEMS die and the PCB. In this manner, the role of mechanically isolating the MEMS from the PCB could be partitioned to the package, instead of the MEMS die, to enable simplification of the MEMS design and fabrication.

Other Functions

The package can also play many other roles. Labels on the package, such as part identification and serial numbers, provide traceability for quality assurance. The packaging enables assembly in non-cleanroom manufacturing facilities, and by methods that would otherwise be harsh to the MEMS die, such as surface mounting and wave soldering. The package's external shape and features could enable assembly by robots.

Finally, packaging can also enhance the performance of a MEMS device. For example, the packaging could provide the MEMS with a high-vacuum, hermetic environment. This is essential to the performance of resonating MEMS devices, where operating in vacuum achieves resonator Q-factors much higher than possible at atmospheric pressure. This type of vacuum encapsulation may be done at either the die or package level, and both options should be weighed when determining to which component to assign that role.

Electronics

Most MEMS require electronics to read out sensors, drive actuators, and interface with the greater product system (see Fig. 9.3). Passive or unpowered devices, such as microstructured surfaces, do not use electronics. Electronics can also add significant value by adding more functions to a MEMS product, such as signal conditioning, calibration, and self-test. Electronics, like the package, often require custom

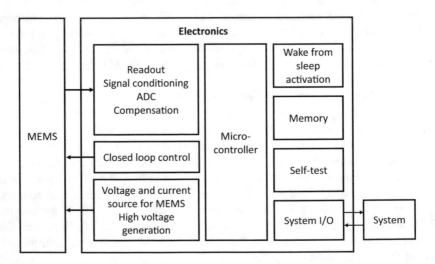

Fig. 9.3 Some of the functions partitioned to the electronics in a MEMS product

design and development in order to meet the needs of both the MEMS device and its end use application. The development of electronics requires significant time and resources and is ideally developed in parallel with both the MEMS and package, as discussed in Chap. 13.

Power

Nearly all MEMS require a power source in order to function. The power subsystem in a MEMS product must provide an electrically quiet, stable source for voltage and current for the MEMS. Noisy power sources can corrupt MEMS sensor or actuator performance.

For actuators, voltages up to hundreds of volts is not unusual. The electronics must generate voltages of this range and apply them accurately in order for the output of the actuator to also be accurate.

Many MEMS sensors are intended to operate in battery-powered systems, where conserving power and maximizing usage time between charges are important product metrics. MEMS electronic design may include power management functions to activate or deactivate ("sleep") a sensor according to when it is needed by the product.

The power draw of the sleep state often determines the overall power consumption of the MEMS product, since in many applications, the product remains in a low-power sleep state most of the time.

An example of how managing the sleep state determines overall power draw is illustrated with the tire pressure monitoring system (TPMS) of an automobile. The TPMS is located in the wheel of the automobile. Due to the wheel's rotation, the TPMS cannot easily be wired for power or communications. Making the electrical connection through the rotating axle would add too much cost. Instead, the TPMS is powered by a battery and communicates with the car wirelessly by radio. The expected lifetime of the TPMS on a single battery is 10 years, so power conservation is crucial. The battery would be quickly depleted if the TPMS system were continuously measuring the pressure. Instead, the TPMS takes advantage of the fact that tire pressure information is only needed when the car is in use. In a daily commute, this is perhaps 1 hour in the morning and 1 hour in the evening (2 of 24 hours in a day). The TPMS can sense that the car is in use through an onboard accelerometer that detects the car's motion. While the car is moving, the TPMS takes a pressure reading once a minute. It may take the TPMS 0.1 s to take the pressure measurement, which means the system is only awake 0.1/60 of a minute, and the rest of the time, it sleeps. Therefore, the overall time that the TPMS system is active is $(2/24 \times 0.1/60) = 0.014\%$ of a day. The remaining 99.986% of the day, the system is asleep. As a result, the power draw of the sleep state dominates its power consumption and therefore determines the overall power consumption of the TPMS.

Readout and Signal Processing

The raw, analog output of a MEMS sensor is not typically usable by the product in which it is incorporated. The signal is often too small and/or noisy. Furthermore, analog signals are easily corrupted. To address these issues, the electronics role is to filter the sensor's signals, amplify the signal to a usable magnitude, and convert the signal from analog to digital form. Once the signal has been digitized, it can be handled robustly and does not require the care that a small, analog signal requires. The sensor signal may then also be calibrated and compensated.

Furthermore, the electronics can add significant value with digital signal compensation. There are a number of nonidealities in real-world sensors and actuators, such as manufacturing variation, zero-bias offset, nonlinearity, temperature compensation, drift, hysteresis, and many more. Some of these nonideal effects can be compensated, at least partially, by signal conditioning methods, which improves the device performance and increases its value.

However, the possibility of digital signal compensation is not license to haphazardly and cavalierly design MEMS devices (and their packages) with nonideal effects. There is a certain amount of reasonable engineering that should be spent on making the MEMS device as ideal as possible, and a certain amount that is more effectively partitioned to signal compensation. The optimal partitioning of addressing nonideal effects between the MEMS and signal compensation will depend on the particular MEMS device, the application, and the nonideal effects. Regardless, the development team should strive to make the MEMS as ideal as cost-effectively possible.

In general, it is good engineering practice to minimize nonideal effects at the source. Compensation has limits. Not all nonideal effects are compensable, and if they are, they cannot be perfectly compensated. In addition, compensation has its own costs and requirements. Compensation capability needs to be developed and integrated into the electronics. Development of compensation requires time and money. Once developed, compensation requires testing and calibrating every unit, which adds unit cost and requires manufacturing test equipment, whose cost can be in the $100 K's to $1 M's. A further discussion on compensation in manufacturing test is in Chap. 20.

Control

Another function of the electronics is to provide control of the MEMS device. For example, when the system sends a request to a MEMS sensor, the request is received by the MEMS control electronics. The control electronics, in turn, execute the necessary actions to take the measurement, and then respond to the system with the measurement, in the correct format, and with the accompanying metadata (such as timestamp, datatype, and error codes).

In addition to this basic functionality, there are a number of other functions that the control system takes care of, such as self-monitoring of the MEMS for health and integrity, power consumption monitoring, updates, and many more. The electronics can also provide closed-loop control of the MEMS device. For example, in resonant MEMS devices, such as gyroscopes, the control electronics keep the sensor in a resonant state, which is required for the sensing operation.

Software

Given the variety of MEMS devices and applications, software may be executed in a programmable component either in the package near the MEMS or outside of the package at the system level. If the software is executed by a microcontroller inside the package, then it is commonly called "firmware." To include software executed either outside or inside the MEMS package, we use the broader term "software" in this discussion.

The presence of software implies a programmable component in the electronics. This is not a required capability in the microsystem, but it can add significant value. For example, higher-level, complex functions may be partitioned to the software because complex algorithms are more cost-effectively embodied in software rather than in hardware. For example, an accelerometer detects a bump, and the output of the accelerometer has a peak. That peak can be identified by either hardware or software methods, but it is typically a larger engineering effort to develop peak identification in hardware. In addition, a software embodiment can do sophisticated peak analyses more cost-effectively than hardware, such as to identify local versus global peaks, measure peak widths, measure peak rise and decay rates, and many other useful analyses of signal peaks.

In certain cases, some functions, like peak identification, may be better partitioned to the electronics hardware while the signal is in analog form. Examples include:

- Peak identification in very high-speed applications.
- The memory space and data bandwidth required to store and analyze many high-resolution samples may be size and/or cost prohibitive.
- The power requirement for a system to work at a high data rate may be impractical.

Software's ability to provide complex functionality enables it to provide even higher-level signal-processing functions, such as event identification. While the raw data coming from a MEMS sensor may be valuable, it would be even more valuable if the MEMS product itself could identify important or triggering events. Doing this signal analysis at the MEMS product level can alleviate the greater system from monitoring a constant stream of data, thereby reducing power, bandwidth, and response time.

An example of effective use of event identification is in wireless systems, where the most power-intensive function is radio communications. A sensor that outputs a constant stream of data by radio will quickly deplete its battery. If instead the sensor is able to identify important and actionable events, it can conserve the radio communications to reporting only those events. Continuing the automotive TPMS example, TPMS must monitor the tire pressure once per minute to check that it is in safe range. By reporting only when the pressure is out of range, the TPMS can reduce communications from once every minute to once every few months or years, or never, depending on how diligently the driver maintains the car's tire pressure.

The flexibility of software also expands capabilities of the development team. For example, a microcontroller present in the product can be programmed to double as a datalogger to monitor performance during development test.

Finally, one of the defining characteristics of software is that it can be updated on a shorter timescale than an update to the hardware. It can take months to develop a fix to an issue in hardware. It can take days or hours to fix an issue in software. A product's failure mode could be temporarily bandaged by a software update, while a more robust, longer-term solution, is executed in the next product generation hardware.

Levels of Integration

Several levels of integration are possible for MEMS products and need to be carefully considered by the development team. Listed in order of increasing integration, the levels of integration are (Fig. 9.4):

- Board level.
- Package level.
- Die level.

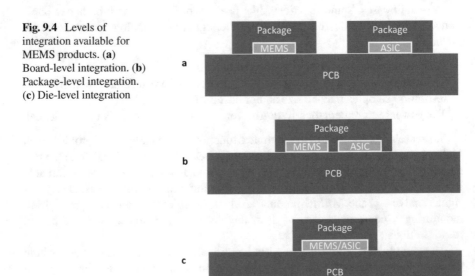

Fig. 9.4 Levels of integration available for MEMS products. (**a**) Board-level integration. (**b**) Package-level integration. (**c**) Die-level integration

Greater integration has potential benefits such as smaller system size, lower power draw, and lower per-unit cost. But to realize these potential benefits, greater integration entails more development cost, time, and risk. The benefits are qualified with "potential" because greater integration does not always guarantee benefits. Many factors and trade-offs must be weighed to decide which level of integration makes the most sense for a particular MEMS device and its application(s).

Board-Level Integration

MEMS products may be most simply integrated at the board level. The "board" is the substrate upon which the system components are gathered and assembled. The components of a board-level system are individually packaged, whether they are MEMS die, microcontrollers, memory chips, resistors, and so on. These individually packaged components are assembled and connected on a printed circuit board (PCB) or flexible substrate, such as a flex circuit.

The board provides a multilayer stack of conductors and insulators that has been patterned to route and connect the individual components in order to create the desired product functions. The typical size of a board-level integrated system is on the order of centimeters to tens of centimeters on a side.

For board-level integrated systems, each of the components are tested and known to be good prior to assembly; as a result, the assembled board is unlikely to have any bad components. The remaining product risk exists in well-controlled and observable assembly steps, such as soldering, so the completed board is likely to yield well.

An example of a board-level integrated MEMS product is an inertial navigation system (INS). The INS requires assembly of multiple packaged MEMS sensors, usually gyroscope, accelerometer, and magnetometer, along with signal conditioning electronics, a microprocessor, power control, and memory modules. The board(s) containing all of these components would then be mounted in centimeter-scale housing, which protects the entire assembly from the environment as well as provides interfaces for external mechanical mounting and electrical cable attach.

Package-Level Integration

Package-level integration consists of bare die components assembled within a single package. A common example is a MEMS die and an application-specific integrated circuit (ASIC) die assembled within a single package. All of the electronics have been integrated into the ASIC die, and the package integrated product contains only two components within the package. The typical size of a package-level integrated product can range from a few millimeters to a few centimeters on a side.

In the board and package-level integrated systems, the MEMS and ASIC die remain two separate die, fabricated independently. Thus, each die may be developed and updated separately. The separation of the MEMS and ASIC die facilitates product differentiation. For example, in an accelerometer product family, accelerometers of different ranges or performance levels could use the same ASIC, but may be differentiated by using different MEMS accelerometers.

Known Good Die and Multichip Packages

Package-level integrated products require assembly of multiple bare die into the package. But most MEMS die can only be fully tested after they have been packaged. This is due to the fact that it is easier for die to be handled, exposed to test stimuli (pressure, acceleration, chemistry, and so on), and electrically read out/driven after the die have been packaged.

Without a complete functional test, it may not be known if the die in a package-level integrated system are good until after package assembly. As a result, the package may be assembled with bad die, and a single failed die will cause the entire product fail. This results in loss of the other good die, as well as the time and cost to assemble and test the package.

This is known as the "Known Good Die" (KGD) problem. KGD compounds combinatorially, which means that overall product yield is a strong function of the number of die in the package and their individual yields. So, for example, if a package-level integrated system contains two die, each with 99% yield from their respective wafers, then the probability that the package integrated system will be good, when assembled from a blind pick of die, is $(0.99 \times 0.99) = 98\%$. The system-level yield drops quickly with increased number of die within a package and with decreased yield from the wafers. For example, if a system contains 2 die with 80% yield from each of their respective wafers, the overall potential yield of the assembled package drops to $(80\% \times 80\%) = 64\%$. If the system contains 3 die each having 80% yield from the wafer, then overall yield of the packaged system drops to $(80\% \times 80\% \times 80\%) = 51\%$. At 51% yield, the per-unit manufacturing cost of the product has effectively doubled.

Given the problem of (un)known good die, it is important to limit the number of die within the package only to those that are absolutely necessary and ensure that the individual die are high yielding.

In addition, test and prescreen the individual die as much as possible prior to packaging. Distinguishing good die from bad is fundamentally a test problem, so if it would be possible to effectively test the unpackaged die, then the known good die issue would be resolved. One possible solution is to test the die while still part of a whole wafer. This strategy works for MEMS that are more fully testable using wafer probe. This includes devices such as resonant oscillators, which can be fully tested through electrical means alone. This also includes MEMS devices for which there are specialized wafer probe heads that can provide physical stimuli to the die at wafer probe. Examples include pressure sensors, microphones, and micromirrors.

See Chaps. 14 and 20 for more discussion on testing and test methods.

Die-Level Integration

Die-level integrated systems assemble the MEMS, the electronics, and sometimes even the packaging, all at the die level. This approach creates the most compact product, measuring as small as a few millimeters on a side.

Die-level integrated systems can offer the lowest per-unit cost embodiment of a MEMS product, but at the expense of high development costs. Given the costs, die-level integrated MEMS microsystems are most often found in high-volume products, where the volume allows the development costs to be recouped.

The die-level integrated development costs are high because of many challenges. One is that the fabrication process of MEMS devices is different than that used for CMOS electronics. These two distinct wafer processes would need to be combined into a single fabrication process flow, which increases the total number of process steps, potentially adding risk and reducing overall wafer yield. Another challenge is that the MEMS and electronics are integrated, so they cannot be independently iterated and upgraded. A third is that the known good die problem cannot be addressed with test methods (more discussion below).

Given these challenges, die-level integrated systems are best for particular circumstances, for example, where the electronics and MEMS must be in very close contact, and/or where the detected signal is small, such as in gyroscopes. Another example is where the number of input/outputs (I/Os) is high, so a large number of interconnects is only practical at the die level. This is the case with the Texas Instruments DLP micromirror arrays, in which there are several million individual micromirror pixels, and several million I/O connections between the MEMS and the ASIC.

An example of die-level integration is the InvenSense (now TDK) gyroscope [5]. It is fabricated on two different wafers, one MEMS wafer and one CMOS wafer, in order to separate their fabrication processes. These wafers are then bonded together, which also creates a die-integrated hermetic vacuum.

Die-level integrated systems, like package integrated systems, suffer from the known good die problem. However, in this case, wafer-level testing would not help to improve yield. For example, in the InvenSense gyroscope, even if a test method could identify good and bad MEMS die on the wafer, the bad MEMS die cannot be removed and they would still be bonded to the good die on the CMOS wafer, and vice versa, reducing overall yield.

Trade-Offs to Consider in Integration

The benefits of integration need to be traded off against practical issues such as more complex fabrication processes, higher development costs and timelines, issues with risk and yield, and product upgrades, to name a few (Table 9.2).

It would be a mistake to assume that more integration always results in a better MEMS product. This misconception derives from semiconductor products, where greater integration often does result in better, cheaper, faster products. (In particular, this gave rise to the System-on-Chip, or SoC approach for digital ASIC design.) For MEMS, however, while there may be performance advantages to increased levels of integration, it should be deployed selectively and strategically. An illustrative example is the Analog Devices' (ADI) ADXL 50, which was a MEMS accelerometer fully die-level-integrated with its electronics in the 1990s [6]. Later, ADI split the MEMS and electronics into two separate chips that were integrated at the package level. Ultimately the package-level integrated system was a better product solution from both technology and cost points of view.

A Common MEMS Product Embodiment

While every application is different and has its own set of integration solutions, there is one embodiment that has been common since the 2000s (see Fig. 9.5), a package-level integrated product that includes the MEMS and the ASIC die. Approximately 50% of MEMS products (in terms of market value) are embodied in this form[7]. The ASIC die is a mixed-signal chip—that is, it contains both digital and analog circuitry, which includes the signal processing, the power conditioning,

Table 9.2 Summary of the trade-offs for different levels of integration

Level of integration	Pro	Con
Board level	Lowest development cost, time, and risk	Largest system
	MEMS and ASIC fabrication are decoupled	Higher per-unit cost
	Independent upgrade of components	Challenges with small signals
Package level	MEMS and ASIC fabrication are decoupled	Known good die yield issues
	Independent upgrade of components	
Die level	Smallest system	Highest development cost, time, and risk
	Potential to offer lowest per-unit cost	MEMS and ASIC fabrication are coupled
	Best for small signals	Known good die yield issues
		Components cannot be independently upgraded.

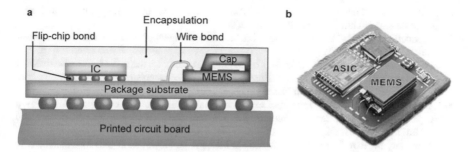

Fig. 9.5 (**a**) A schematic of a package-integrated MEMS product, which includes the MEMS, ASIC (IC) and package. (**b**) Photo of uncapped packaged integrated microsystem and assembled bare die. (Reprinted with permission from Fischer et al. 2015) [7]

the control electronics, and the digital compensation. The ASIC also includes a microcontroller with memory to store and run software.

The package-level integrated embodiment is favored for a number of reasons. It provides the essential elements of the MEMS, package, and electronics, and can also capture additional value provided by the software. Furthermore, package-level integration minimizes system size and per-unit costs, while avoiding the development and yield difficulties associated with die-level integration. The MEMS and ASIC are separately fabricated to avoid integration of their two different fabrication processes. Finally, the ASIC and MEMS can be separately updated to provide the flexibility to create a family of products by swapping different MEMS and ASICs.

Summary

The MEMS product is a system, which is typically composed of the MEMS device, package, electronics, and software. Understanding the functions of each component of the system allows the development team to increase the overall value of the product and to partition functions to the appropriate components. The MEMS product can be integrated at board, package, or die levels. Greater integration can create significant advantages, such as smaller products and lower per-unit costs, but also has greater challenges, such as higher development cost, time, and risk in order to realize those advantages.

References

1. Maboudian R, Ashurst W, Carraro C (2000) Self-assembled monolayers as anti-stiction coatings for MEMS: characteristics and recent developments. Sensors Actuators A Phys 82(1–3):219–223

2. (2003) Sound Power. Engineering ToolBox. https://www.engineeringtoolbox.com/sound-power-level-d_58.html. Accessed 31 May 2020
3. Introduction to Solar Radiation. Newport. https://www.newport.com/t/introduction-to-solar-radiation. Accessed 31 May 2020
4. Ratanawilai T, Hunter B, Subbarayan G et al (2003) A study on the variation of effective CTE of printed circuit boards through a validated comparison between strain gages and Moire interferometry. IEEE Trans Compon Packag Technol 26(4):712–718
5. Seeger J et al (2010) Development of high-performance high-volume consumer MEMS gyroscopes. In: Solid-State Sensors, Actuators, and Microsystems Workshop, Hilton Head Island, SC, June 2010 pp 61–64.
6. Mehregany M, Roy S (1999) Introduction to MEMS. In: Helvajian H (ed) Microengineering aerospace systems, the aerospace press. El Segundo, California, pp 16–18
7. Andreas F, Forsberg F, Lapisa et al (2015) Integrating MEMS and ICs. Microsyst Nanoeng 1(1):1–16

Chapter 10
Starting a New MEMS Device Design

MEMS design involves uniting several physics domains and engineering disciplines into one cohesive design that is best regarded as a system, as previously described in Chap. 9. Significantly, a MEMS chip design is conjoined with its process flow; you cannot design one without carefully considering impacts to the other, and vice versa. Additional requirements are created by the MEMS chip's electronics, packaging, operating environment, field reliability and how each of those interact with the physical chip design and vice versa. When starting a MEMS product design from scratch, it may feel daunting to find a place to begin the development work when considering these multiple complexities and interactions.

In this chapter, we present a methodology for how to *begin* a new MEMS design. We emphasize the word "begin" deliberately. Depending on the type of product, whether sensor, actuator, microfluidics, or passive microstructure, and its intended end use application, the actual pathways that will bring the product design and development to completion will be quite different. For all that variety, however, they all start from the same origin, which we describe here.

Overview of New Product Design

Chapters 10–14 introduce the main tasks of developing a novel MEMS product. The sequence in which we chose to present these tasks occurred simply because in printed material, chapters need to be linearly sequenced. The chapters, however, might be better imagined as jigsaw puzzle pieces, each of which interlocks with multiple other chapters. The manner in which the product design and development work will actually get done is not a linear sequence; there will always be parallel efforts and looping iterations as development proceeds.

Chapters 10–14 outline the interdependent and parallel tasks of designing the MEMS device, its fabrication process, back end and packaging steps, electronics and related components, and test features, respectively. Throughout the product development, you will need to iterate frequently among each of these critical puzzle pieces, making adjustments and trade-offs as you proceed in order to get each of them to fit well together.

Eventually, you will not so much finish the overall product design as converge it to a solution point at which all of the key pieces that comprise the product more or less fit together. In MEMS, it is usually possible to find multiple workable solutions, one of which must be selected according to your business's appetite for risk, cost, and time to market.

In this chapter, we describe the starting point of a new product design, which is to design the component at the heart of the product: the MEMS device.

The main steps for beginning a new MEMS device design from scratch are:

- Explore design space using analytical models and then down-select to a target space based on practical considerations.
- Develop concept design(s) compatible with the target space and build simple computational models of them.
- Enhance models to include more complexity.
- Consider inputs from other system components or physical domains.
- Manage model uncertainties; if possible, with test and measurement.
- Iterate.

If you are an inventor who has been developing a new technology and may already have built a proof-of-concept prototype, such as during Ph.D. graduate research, it is still worthwhile to start fresh, following these design steps. Proof-of-concept prototypes made during research programs are created for the purpose of validating device physics and viability of a novel technology; these are not, and were never intended as, product designs or prototypes. Fully exploring the potential design space is the first step towards designing a product.

Explore the Design Space Using Analytical Models

When starting a new design, you may feel a temptation to draw your MEMS idea in 3D CAD software and then import it to a computational software package right away, due to an understandable desire to quickly visualize and understand your design. All of the advanced computational modeling software available today, whether based on finite element, finite volume, or finite difference computational methods, certainly make it very easy to indulge this temptation.

You will, however, gain more value and more insight, much faster, by first starting to build a basic analytical model of the device physics within a spreadsheet or an analytical software package like MATLAB.

By basic analytical model, we mean the set of equations or transfer functions, which describe the stimulus (or stimuli) to be input to the MEMS device, and what outputs will result from the transducer's physical response.

Table 10.1 Recommended parameter analyses for some common MEMS device types

MEMS device type	Design parameters to explore
Pressure sensor	Deflection versus pressure, membrane stress
Microphone	Deflection versus pressure, membrane stress, resonant frequency
Comb drive (inertial sensors, actuators)	Electrostatic force versus finger length, spacing, and input voltage
Resonators	Resonant frequency, modes, Q factor
Microspeaker or Ultrasound transducer	Deflection versus drive voltage, membrane stress, resonant frequency
Piezoelectric actuator	Displacement or force versus input voltage
Isolated thermal platform (calorimeter, bolometer, gas sensor)	Thermal conductivity versus geometry; heat loss versus time
Cantilever	Deflection versus force or pressure, root stress
Valve (gas or liquid)	Spring constant, sealing or opening force

At this stage, the analytical model need only be accurate only to first order. Its main purpose is to validate that the basic transducer function appears physically possible and that the geometry and characteristics required to achieve the desired performance are indeed practical and achievable at the MEMS scale.

It may not be as fun to kick off your design efforts by developing a set of differential equations; however, it will be far more effective than beginning with 3D doodles of designs, for reasons we will soon make clear. Fortunately, many excellent textbooks and sources provide well-established analytical models to help jump-start your efforts [1–18]. Table 10.1 lists some common MEMS device types and their respective key design parameters to model.

Spreadsheets or MATLAB are particularly useful tools for analytical modeling because they can quickly generate X-Y graphs (as well as more complex graphs) for a wide range of design input values and their respective outputs. In contrast, rendering each of these design inputs as multiple individual CAD designs and then running computational simulations of each would be painfully time-consuming and make data synthesis more difficult. Graphs of analytical model data can be quickly assessed by eye and by engineering judgment, and then segmented into regions of ideal, workable, or not useful solutions. The popular term for generating these plots, locating sweet spots of favorable interaction between variables, as well as no-go regions, is "exploring the design space."

Using Design Space Plots

Design space plots efficiently illustrate where you have the highest probability of finding a workable solution, and ultimately a successful product design. A large design space is highly desirable at the beginning of any product design effort, but is particularly important in MEMS. Early in MEMS development, it provides much-needed maneuvering room for the inevitable trade-offs with other important, yet-to-be-designed system components, such as the package and electronics, as well as fabrication conditions, all to be discussed in Chapters 11–15.

If an analytical model were to reveal that only a narrow range of input variables can achieve the desired device output, then the design space is small. Small design spaces can unfairly burden other aspects of the overall product design, for example, by requiring very tight manufacturing tolerances or having electronics function within a narrow voltage range (See illustrations in Fig. 10.1).

While small design spaces may still offer workable solutions, they generally require more finesse, skill, and cost to execute, especially once all other aspects of the product design are considered. If an analytical model were to show that a concept's design space is nil, then abandon the design, and go back to the drawing board for other ideas.

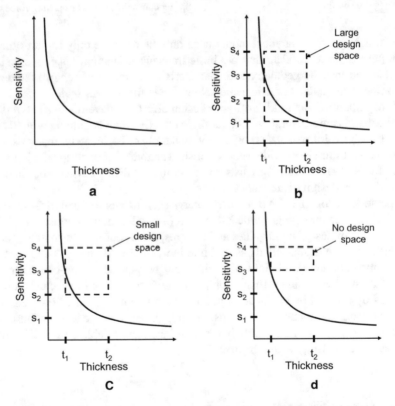

Fig. 10.1 Design space example. Imagine a sensor having its sensitivity as a function of a specific feature thickness, as illustrated in (**a**). The feature thickness, t, is constrained by manufacturing capability to $t_1 < t < t_2$. For a given sensor application, if an acceptable sensitivity, s, is $s_1 < s < s_4$, then the curve falling within the design space box indicates that any thickness, $t_1 < t < t_2$, will be suitable, so the design space is large (**b**). If an acceptable sensitivity were $s_2 < s < s_4$, then only a narrow range of thicknesses will be suitable and the design space is much smaller (**c**). If an application were to require sensitivity $s_3 < s < s_4$, then there is no design space within the thickness constraints, so the entire design must be reconsidered (**d**).

Once you understand and identify the limits of a device's design space, you can then confidently invest more time and resources on detailed, higher accuracy models. The analytical model enables thoughtful down-selection to a narrower set of input variables, or target design, upon which to focus computational modeling efforts.

Additionally, an analytical model is also extremely useful to the development team's electronics designers. At this early stage, an assessment of the electronics' design space is also needed. For example, would the voltage required to drive the MEMS device be realistic for the end use application? Is the voltage sustainable by the envisioned power source, such as a battery, solar cell, or inductive power? Co-design and co-modeling of a MEMS device and its electronics begins at this early stage. For more information, please see Chap. 13.

Mind the Practical Limits of Analytical Models

Although initially quite useful for assessing design space, analytical models do have practical limits for developing a device design. When working with an analytical model, be aware and respectful of any underlying assumptions in your system of equations. Refer to fundamental textbooks of physics, mechanical, and electrical engineering in order to refresh your memory on which assumptions may underpin your calculus.

Beware the effects of simplifying assumptions in analytical models, such as:

- Elimination of higher order terms in differential equations
- Geometric length-to-width ratio limits, such as in beam theory
- Plane stress or plane strain
- Infinite plane
- Neglect of parasitic effects, such as in capacitive calculations
- Perfect structural boundary conditions, which may not be quite so perfect in a real MEMS device
- Fluid dynamic regimes based on figures of merit, such as Reynolds or Knudsen numbers, that may be vastly different at microscale
- Neglect of temperature dependent material behaviors

As a design concept evolves and begins to violate simplifying assumptions in the analytical model, it is time to move on to computational modeling.

Building Useful and Meaningful Computational Models

Having completed and explored an analytical model, you will be well equipped to begin building an accurate and useful computational model. Depending on the physics involved, you may need to use finite elements (most commonly used for solid structures) or finite volume method (used for fluid dynamics) or finite difference method (used for electromagnetics). We will focus this discussion on multi-physics finite element (FE) software, because it is most commonly used for MEMS design work.

There are three key steps to getting the most value from a finite element model, which hold true whether working on a model of a macro- or microscale device:

- First, build a simple FE model and then use the analytical model or other hand calculations to validate it.
- As the design and understanding of its physics and end use application develops, refine the FE model.
- To get the most accurate results, invest in validating the model using measured data from actual devices and plan to continually fine-tune it over the life span of the product.

Build a Simple Model First and Validate It

An analytical model provides insights on your design and therefore will help you to better define the scope of a FE model. Most significantly, the analytical model guides you on which geometric features must be included in the initial FE model and which features may be safely omitted. For example, you may realize that although simple beam equation calculations were no longer valid due to aspect ratio and expected large deflections of your device, a two-dimensional FE model including nonlinear behavior would be valid and sufficient to answer your initial design questions. You could thereby temporarily avoid creating a featured 3D model, which takes more time to build and run.

Although today's FE software and computer capabilities facilitate building a FE model straight from a fully featured 3D CAD solid model file, starting with a fully featured model could slow you down. A common challenge, for example, is mapping a finite element mesh to relatively small features on a solid model. Small features require small finite elements that must then be propagated into the rest of the solid model, adding many more finite elements to the model that would not otherwise be needed. Larger numbers of elements increase computation time. If the small features are unimportant to understanding the overall device physics, it is best to get rid of them, a process known as "de-featuring" a model. Particularly when working on simulations such as transient response, nonlinear, or contact between bodies, having a simpler FE model containing fewer elements will make it much easier and faster to converge the model and get results.

As you gain insight on the physical behavior of your device, its end use application, and add more detail to your design, you will need to further enhance the FE model. There should be a progression in the complexity of the model that parallels the evolving MEMS design (see Table 10.2). A second iteration of a FE model, for example, might then expand the model from 2D to 3D, or revise the Young's modulus of silicon from a simplified isotropic value to the full anisotropic tensor.

Table 10.2 As MEMS development proceeds, the complexity of a model must evolve to provide richer and more accurate predictions

Model characteristic	Phase of development			
	Concept development	Detailed design work	Prototype	Mature design
Dimensionality	2D	Simple 3D	Fully featured 3D	Fully featured 3D
Material properties	Bulk material textbook values, isotropic	Bulk material textbook values, anisotropic and oriented to crystal plane (as needed)	Measurements of as-deposited material properties	Data lookup table including material property variation with temperature, aging effects, etc.
Boundary conditions	Idealized	More realistic	Adjusted based on measurements on prototype devices	Adjusted based on measurements on prototype devices
Residual stress	Expected thin film stress	Measured thin film stress from test wafers	Measured thin film stress from prototype wafers	Thin film stress data with 2σ tolerance values
Electronic models	Calculated parameters such as voltage, current and power	MATLAB, Simulink and SPICE models	EDA and ASIC development models	Models for complex behavior such as noise, phase jitter, parasitics, etc.
Typical accuracy	Coarse	Within 15–30% of reality	Within 5–15% of reality	Within 1–5% of reality

As illustrated in the last row of Table 10.2, throughout the design development and FE model evolution, you must frankly assess the likely accuracy of a model's results and keep that in mind when making design choices and trade-offs. Early-stage models will likely be accurate to within only 15–30% of real-world results because of uncertainty in important items such as thin film material properties, etch tool performance, and so on. Fortunately, it is possible to mindfully manage these uncertainties and still get useful results from a FE model, described in more detail below.

Incorporating Input from Other System Components

So far, our discussion has focused on modeling of the MEMS device itself. As stated earlier, the MEMS device is one component that works in concert with electronics, a package, and other components in order to provide a transducer function.

After a preliminary model of the MEMS device has been created and validated, additional inputs to the model must be examined, such as:

- Dimensional tolerances or practical constraints due to fabrication process
- Electrical signal input
- Thermal input, whether from environment or adjacent components

- Stress from chip-mounting method and packaging
- Environmental inputs, expected or spurious
- Location and type of mechanical, electrical, optical, and/or fluidic interfaces
- And many more, specific to your product

Please see Chaps. 11–15 for more details on other aspects of MEMS system design that may need to be incorporated into your model.

Throughout FE model evolution, information must be continuously shared with the electronics and packaging design teams, who will be building in parallel their own models of the readout, control or power electronics and packaging conditions. Some MEMS-specific modeling software can pass design parameters between multi-physics simulation environments and circuit simulation environments, in order to facilitate parallel development of the device, its electronics, and its package.

For MEMS devices that function in closed-loop electronics control, such as gyroscopes and oscillators, co-development of the mechanical and electrical domain models is essential. One model does not pass down requirements to the other; the two models must mutually converge to a solution point through iteration.

Over the course of MEMS product development, the device and component designs, and all of their models, must be continuously updated to reflect growing knowledge about materials, fabrication processes, electronics, environment, etc., and all of their complex interactions. Revisit these evolving design inputs frequently and then iterate, iterate, iterate.

Example of the Benefits of Simulation: Understanding Environment Impact on Device Performance

MEMS sensors and actuators are susceptible to stray and corrupting environmental inputs which may distort function in real time or may actually damage the product. Most MEMS are susceptible to package-induced stress (see Chap. 13), temperature, and may also be susceptible to shock, vibration, orientation with respect to Earth's gravity, electromagnetic interference, and so on.

Testing the effects of these environmental inputs requires specialized and expensive equipment, such as thermal chambers, vibrational shakers, or rate tables. There is also significant time and expense involved to set up and run physical tests, and then to analyze all of the gathered data.

One major benefit of having an accurate FE model is using it to analyze the effects of environmental inputs on device performance. Within the modeling environment, it is straightforward to apply many different types of environmental inputs, singly, or superimposed, and to assess their impact on device function. Modeling studies can very efficiently assess and filter the significance of various environmental inputs, thereby enabling intelligent prioritization of expensive environmental testing for only on the most significant effects or for those required for product quality certifications (see Chap. 14).

Managing Uncertainty in Models

One of the challenges of modeling MEMS is dealing with uncertainty in design input variables. The typical sources of uncertainty in MEMS models are:

- Material properties
- Structural boundary conditions
- Fabrication tool performance, uniformity, and tolerance
- Micro- or nanoscale phenomena

These sources of uncertainty, and how to deal with them, are described in more detail below.

Sources of Uncertainty in Modeling

MEMS are made from materials which are grown or deposited, nanometers at a time, by chemical or physical processes having multiple input variables. As a result, there is often uncertainty in the exact composition and properties of MEMS materials, as well as in the repeatability of those items from tool to tool or run to run. In addition, thin films will usually have a residual stress resulting from coefficient of thermal expansion (CTE) mismatch and/or lattice mismatch with the underlying material upon which it has been deposited. Thicker films (1–10 microns) may also have stress gradients through the thickness of the film.

Having accurate material properties data for a MEMS model requires investment in time and resources in order to measure the materials, made by a specific set of tools, ideally those which will be used in production. In addition, once measured, the validity of those material properties will depend on maintaining careful control of the processes over time, particularly for thin films. Any significant deviation in tool performance may cause a shift in material properties. Methods for measuring MEMS-scale material properties have been developed and are described in the literature [6]. Until you have the means to directly measure the properties of materials made by your fab's tools, you will need to use estimated values, gleaned from textbooks or articles, in your FE models.[9, 10, 17]

The boundary conditions of MEMS structures also pose a challenge due to uncertainty in actual microscale conditions. For example, in modeling a macroscale mechanical design such as a bridge, a cantilever boundary condition provides a high fidelity representation of the real structure. In the macro world, it is possible to build a cantilever having a root attachment so rigid that displacements are effectively zero (Fig. 10.2a). In a thin film MEMS cantilever suspended over an etched silicon cavity, however, the root of the cantilever is not truly fixed because its top surface is free to stretch and deform (Fig. 10.2b). Nor does the MEMS cantilever have a simply supported boundary, either. A more realistic boundary condition for modeling a thin film MEMS cantilever will be somewhere in between simply supported and cantilever; where exactly will depend on the design and fabrication particulars.

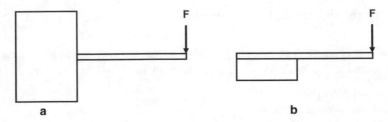

Fig. 10.2 Boundary condition example: cantilever beams with a tip load, shown in cross section. In a macroscale cantilever (**a**), a large support can be built around the cantilever, creating an idealized cantilever support with no root displacement or rotation. In a MEMS device (**b**), a microscale cantilever structure may be made from a thin film suspended over an etched cavity. The top surface of the MEMS cantilever is free to displace and rotate, whereas its bottom surface is fixed to the substrate. The true boundary condition for cantilever (**b**) is not the same as for (**a**).

Fabrication tool performance also adds uncertainty to the accuracy of a FE model. Within a model, it is quite easy to draw perfect shapes representing an idealized MEMS structure. Yet in MEMS fabrication, due to multivariable process conditions, it may be quite difficult to fabricate those shapes to the same precision as drawn, let alone with any consistency within a chip or across a wafer. Due to these manufacturing variations, the MEMS performance could turn out to be quite different from what the model predicted.

This issue commonly occurs in MEMS made by deep reactive ion etching (DRIE). The tool's etch performance is highly sensitive to many conditions, such as type of pattern masking material, overall exposed silicon area on the area, feature width, shape and aspect ratio, location on the wafer, to name a few. The etch tool will not be able to create the perfect shapes drawn in a FE model (Fig. 10.3). For some types of MEMS devices, small deviations in the etch pattern will have large effects on performance. In these situations, it will be necessary to update and iterate the FE model to account for measured etch performance in order for the model to produce usefully predictive results.

Another challenge for MEMS is accurate modeling of micro- and nanoscale phenomena. The tiny dimensions of MEMS create physical interactions which are not present, or are inconsequential, at the macroscale (millimeters and larger). One example is the microscale nonlinear damping behavior known as squeeze-film damping that occurs from viscous fluid motion between two microscale moving objects. While a basic physical model of the phenomena has been implemented in FE tools such as ANSYS, it cannot capture the full range of potential MEMS device behaviors, particularly at higher frequency.

Attraction forces between microscale objects in close proximity or that touch lead to "stiction," a usually irreversible welding of two surfaces that results in device failure. A number of different physical phenomena influence stiction behavior, such as liquid capillary, Casimir, or van der Waal forces. It may not always be obvious a priori which particular forces are at play. Furthermore, specific surface states resulting from fabrication conditions, such as asperities or dangling chemical bonds, also influence stiction behavior. For this reason, it is still difficult to model stiction

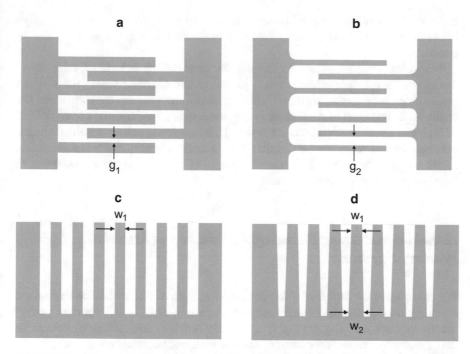

Fig. 10.3 Ideal shapes used in a model may not be achievable by etch fabrication methods. Model plan view of an electrostatic comb drive, designed to have gap, g_1 (**a**). Realistic outcome from DRIE, in which lateral etching and rounding occurs, creating a larger gap, g_2, which will have a higher actuation voltage than g_1 (**b**). Model cross-section view of gyroscope features, designed to have uniform width of w_1 (**c**). Realistic outcome from DRIE, in which tapering occurs, resulting in a non-uniform cross section of w_1 at the top and w_2 at the bottom. This type of etch result will cause a quadrature error in gyroscope performance.

behavior in a deterministic manner. Next-generation product design efforts could implement an empirical model of the behavior, derived from observation and measurement of stiction in functioning MEMS devices from an earlier generation product, in order to fine-tune the FE model.

Finally, it is worth briefly noting that computational models can also have uncertainties resulting from the numerical methods in use. In FE, for example, computational uncertainties may arise from inadequately meshed features or user error in choice of element type.

Parameter Sensitivity Analysis

A model is a tool to extract deeper insights on the physical behavior of a design. A lot may be learned from observing how model output changes in a relative manner as inputs are systematically varied, a practice known as parameter sensitivity analysis. The relative changes in output due to varying inputs contain useful design information even if the absolute value of the output is not too accurate due to uncertainty.

Managing uncertainty means keeping in mind which aspects of your initial model results are approximate versus accurate, and knowing where the model's assumptions may break down and where uncertainties may strongly influence the result. To do this level of interpretation requires a strong fundamental understanding of the basic physics underpinning a device. If, for example, you are a trained electrical engineer and you are not so familiar with structural behavior, this is a good time to ask a colleague trained in mechanical engineering to check your model and to help you correctly interpret its results.

Example: Managing Boundary Condition and Material Property Uncertainties

For example, in the design of a simple resonator, such as a beam having fixed-fixed boundary conditions, the natural resonant frequency, or f_o, is an important performance requirement. Calculating f_o accurately requires accurate input values for beam geometry, Young's modulus of the material, and boundary conditions.

At the start of the design process, you may only know the geometry accurately, and not know the actual Young's (elastic) modulus of the thin film material from which it will be made, nor the precise state of the boundary conditions at each end of the thin beam.[1]

Create a first model using the Young's modulus of the material in bulk form (such as may be found in a textbook or reference manual) and approximate the boundary conditions of the beam as fixed-fixed (zero displacements and zero rotation at each end of the beam). In this first model, your goal is to understand how the f_o varies as the beam's length, thickness, and shape varies. As you vary these inputs one by one, while holding the Young's modulus constant, examine the *relative* changes between the f_o outputs.

The absolute values of f_o at each design point calculated by your model will be approximate because the Young's modulus is not quite right, nor are the boundary conditions. Due to the use of fixed-fixed conditions on a thin film beam, we know that the calculated absolute value of f_o will be higher than in reality, because a thin film beam will have more flexible boundary conditions that would result in a lower f_o. However, the results on f_o for each of the varied design inputs will still provide useful information about the *relative* effects of varying the beam geometry. For example, if the beam thickness increases by $x\%$, the f_o will increase by $y\%$. You may use these relative results to gain insights on how different input parameters affect the performance of your design.

---

[1] Thin films suspended over the edge of an etched silicon cavity do not have the same boundary conditions as the classical fixed-fixed beam if the top side of the film is unconstrained.

Use Measured Data to Improve Model Accuracy

Understanding where uncertainties exist in a FE model can usefully inform the selection of test structures, experiments for fabrication short loops, or early proto-typing runs (see Chap. 15). Any missing or approximated model input data, such as material properties, should be harvested by measuring as-fabricated materials and devices. Models created in commercial FEA software can usually be enhanced with custom programming to include, for example, lookup data tables or polynomial fits from previously measured materials data.

Measured data from functioning devices should be cross-checked against model outputs, and any observed discrepancies need to be investigated and resolved. Continuing the example shown in Fig. 10.2a, b, if testing of an electrostatic comb drive prototype revealed higher than expected actuation voltage, then at minimum, the comb tines should be visually inspected and measured. If the etched comb tines turned out narrower than expected, then either the etch recipe will need to be adjusted and rerun in order to achieve a result as close to Fig. 10.2a as possible, or if the condition of Fig. 10.2b is actually acceptable, then the FE model must be updated to better reflect the realistic etch conditions.

Iterative measurement and FE model tuning takes time and resources, and only engineering judgment can determine what is "accurate enough." A properly tuned model that provides accurate results is a powerful and valuable tool for next-generation product development. By accurately predicting the performance of new design variations, a tuned model could enable design validation to occur inexpensively and quickly on a computer.

When Testing and Measurement are More Efficient or Useful than Simulation

There are situations in which simulation cannot provide immediate reasonable answers and you must do testing and measurements on prototypes in order to inform your continued product development.

Examples of phenomena which cannot be easily modeled and therefore need direct measurement are:

- Microscale material time-dependent phenomena, such as fatigue or creep
- Multivariable environmental effects, such as corrosion
- Sensor output drift
- Estimating mean time to failure (MTTF) of a device

Many materials exhibit time-dependent behaviors, such as fatigue, creep, and aging. Physical and environmental inputs over time, such as cyclic stress, constant stress, humidity, thermal cycles, or even light exposure, may cause degradation in a material (particularly its mechanical strength) over time. Thin-film metals and polymers are especially susceptible to time- and temperature-dependent behaviors, and polymers are even sensitive to UV light exposure.

FE modeling requires a deterministic, mathematical relationship which describes the phenomena as a function of various input conditions; if no math exists because the physics of the phenomena are not well understood, then only empirical methods may be employed in models. While empirical models exist for certain behaviors as they occur in bulk or macroscale materials (for example, fatigue models for behavior of aluminum alloys or solders), there are very few models available on such behaviors in microscale materials.

Test and measurement provide an efficient means to address these uncertainties. Once a body of test data has been accumulated by observing and measuring a phenomenon in actual devices, then it may be possible to create an empirical model based on that data. Empirical models may then be included in FE models with additional programming such as use of lookup data tables or polynomial curve fits. One such example is a silicon fracture prediction model that we created to estimate the probability of fracture under load as a function of etched silicon surface roughness [19, 20].

Business Factors in MEMS Design

In our business, we have seen many customers, especially those at start-up companies, resist using modeling and simulation in early stages of MEMS product development. A common reason is the organizational need to show some physical item or prototype to investors or executives as a convincing demonstration of development progress.

Since investors and executives usually do not have the expertise to critically evaluate modeling results, they remain rightfully skeptical of simulation's pretty rainbow pictures. Models can be easily botched or manipulated, whereas a functioning hardware prototype is a reliable demonstration of feasibility. It is no wonder that investors and executives usually demand to see working prototypes before committing more funding. As a result, entrepreneurs and inventors often focus their limited time and development budget on building prototypes, not models.

In contrast, in today's practice of macroscale mechanical and electromechanical product development, modeling is a mandatory step in the design process. To skip it would be unthinkable, because modeling and simulation of macroscale technology, such as aircraft frames, can validate designs with high accuracy. Standardization of commodity bulk materials (metal alloys, in particular) and extensive characterization of material properties, combined with a rich understanding of mechanics and material behaviors accumulated over centuries of mechanical engineering, make FE simulation a powerful and essential design tool. Companies devote teams of engineers solely to simulation because its value to the development effort is so clear. Highly experienced MEMS design teams and companies similarly understand the value of simulation, invest resources in modeling, and use it to their advantage.

Yet among MEMS innovators and entrepreneurs, we have seen simulation treated as an afterthought. Whether due to unfamiliarity with FE modeling (which is sometimes the case for those who are electrical engineers), concerns about uncertainties in model inputs, or under pressure from investors or executives to deliver a working prototype, the urge to run into the fab and start fabricating a MEMS design is nearly irresistible.

Leverage Simulation to Reduce the Cost, Risk, and Time of Product Development

In MEMS product development, there is a stiff penalty for building an ill-considered prototype that ends up in the silicon recycling bin.

We can estimate the impact of a failed wafer fab run on schedule and budget. As an example, imagine developing a 6-mask layer MEMS device on 150 mm wafers, with plans to run a prototype lot of 10 wafers at an R&D facility. Assume the burdened labor cost (i.e., including overhead, general and administrative expenses) for a fab engineer or operator is US$100 per hour. Creating the mask data, including the wafer layout, alignment marks, test structures, as well as several design variants, takes one person-month, or 160 hours. The cost of processing one mask level in the MEMS process is estimated at US$15,000 per level, including the cost of the photomask itself, and takes one week. For this example, the total cost of preparing for the prototype run, processing the wafers, and then testing the finished wafers is shown in Table 10.3.

Based on these simple assumptions (rerun the calculation with your own labor rates and fab expenses), we estimate that one R&D run will cost US $114,000 and take 9 weeks.

For a company of any size, this is a significant investment in time and money. For a start-up company, this sum could be the equivalent of one engineer's annual salary. No one would want to needlessly consume such a large amount of money on a batch of wafers that yields no working devices nor any useful data due to an error in design.

This example provides a modest estimate, corresponding to a low complexity MEMS device such as a pressure sensor. Many MEMS devices rely on custom

Table 10.3 Estimated cost to prepare, process, and test wafers from one development run. All costs in US dollars. The reader should update this example using costs typical in their company and geographic region.

Operation	Person-hours	Cost	Time (weeks)
Mask layout and checking	160	$16,000	2
Fabrication cost		$90,000	6
Test and measurement	80	$8,000	1
Total		**$114,000**	**9**

wafer substrates, having order lead times that could add at least 8 weeks to the time-line; expensive specialty steps such as CMP and wafer bond, or stealth dicing; and many more photomask layers. Fabricating a more complex MEMS sensor, such as a gyroscope, could consume at least twice as much money and time as shown in this example. The need to prevent failed fabrication runs gets more urgent with increasing device complexity.

To continue this example, if solely relying on fabricating prototypes to learn about and improve a design, an implication of Table 10.3 is that there would only be a maximum of five learning cycles within a calendar year, at a cost of at least US $570,000. Due to the need for redesign time and extensive testing between fab runs, more realistically, only two to three cycles of learning would be possible within a calendar year.

In other words, relying solely on design-fab-test cycles to advance a MEMS design is an extremely slow and expensive way to do product development.

The conclusions from this simple example are:

- Failed fab runs waste significant time and money
- Cycles of learning from fab runs are slow and expensive

If a model could prevent just one failed fab run, it will more than pay for the engineer's time and effort to set it up. If it could prevent several failed fab runs, well then, it will also pay for the cost of the FEA software as well as the annual salary of an engineer to build the model.

Most significantly, modeling speeds up cycles of learning by enabling design experiments to be quickly and inexpensively conducted on a computer, not in the fab. Finally, models can become a valuable repository in which to accumulate deep knowledge about a design, and then be powerfully applied to achieve efficient and accurate design of next-generation products.

Summary

In this chapter, we described how to begin work on a novel MEMS design by first using an analytical model to explore the design space, followed by constructing computational models to provide more detailed and thorough analyses. Even though several uncertainties can plague MEMS models, these can be overcome with intelligent interpretation of results and diligent test and measurement in order to fine-tune and validate the model with empirical data. Models can be used to efficiently explore many design variables that would be too cost- and time-prohibitive to build in prototypes.

The following four chapters will describe additional key components, all of which must be developed in parallel when creating a new MEMS product. Iterate among them until all the puzzle pieces fit together in a workable and cost-effective design.

References

1. Adams TM, Layton RA (2010) Introductory MEMS: fabrication and applications. Springer, New York
2. Allen JJ (2005) Micro electro mechanical system design. CRC Press, Boca Raton, FL
3. Baltes H, Brand O, Fedder GK, Hierold C, Korvink JG, Tabata O (eds) (2005) CMOS-MEMS: advanced micro and nanosystems, vol 2. Wiley-VCH Verlag, Weinheim
4. Bhugra H, Piazza G (2017) Piezoelectric MEMS resonators. Springer, New York
5. Bryzek J, Petersen K, Mallon JR, Christel L, Pourahmadi F (1990) Silicon Sensors and Microstructures. Lucas NovaSensor, Fremont
6. Cassard JM, Geist J, Vorburger TV, Read DT, Gaitan M, Seiler DG *Standard Reference Materials® User's Guide for RM 8096 and 8097: The MEMS 5-in-1*, 2013 Edition. National Institute of Standards and Technology Special Publication 260-177
7. Doll JC, Pruitt BL (2013) Piezoresistor Design and Applications. Springer, New York
8. Esashi M (2011) MEMS (in Japanese). Morikita, Tokyo
9. Ghodssi R, Lin P (eds) (2011) MEMS materials and process reference handbook. Springer, New York
10. Huff M (2020) Process variations in microsystems manufacturing. Springer, New York
11. Korving J, Paul O (2006) MEMS: a practical guide of design, analysis, and applications. Springer, New York
12. Kovacs GT (1998) Micromachined transducers sourcebook. McGraw-Hill, New York
13. Madou M (2011) Fundamentals of microfabrication and nanotechnology, 3rd edn. CRC Press, Boca Raton
14. Maluf N, Williams K (2004) An introduction to microelectromechanical systems engineering, 2nd edn. Artech House, Boston
15. Senturia SD (2004) Microsystem design, 2nd edn. Springer, New York
16. Sze SM (ed) (1994) Semiconductor Sensors. John Wiley & Sons, New York
17. Tilli M, Paulasto-Korchel M, Petzold M, Theuss H, Motooka T, Lindroos V (eds) (2020) Handbook of silicon based MEMS materials and technologies. Elsevier, Amsterdam
18. Young WC (ed) (1989) Roark's formulas for stress and strain, 6th edn. McGraw-Hill, New York
19. Fitzgerald AM, Pierce DM, Huigens BM, White CD (2009) "A General Methodology to Predict the Reliability of Single-Crystal Silicon MEMS Devices." Journal of Microelectromechanical Systems, 18(4), 962–970. doi: 10.1109/JMEMS.2009.2020467.
20. Fitzgerald AM, Pierce DM, "Fracture prediction for crystalline microstructures" U.S. Patent 7,979,237

Chapter 11
Design for Manufacturing: Process Integration and Photomask Layout

MEMS process integration, the development of a compatible, reliable, and cost-effective process flow to bring your device into reality, remains a critical and often problematic bottleneck in MEMS development. What appears to work beautifully on paper unfortunately does not always work well in practice due to intricacies in how the layout and different process steps interact with each other. Process integration means to determine a sequence for fabricating your MEMS device, process step by process step, while making sure that each step will not damage prior steps nor cause issues for subsequent process steps. With all of the variables at play in silicon processing, good process integration requires careful thought and attention to detail.

Having awareness of whether your full MEMS process flow is realistic or aggressive, unique or common, and of all the associated trade-offs, can help you to determine the best path forward in development. For instance, if your process flow requires a novel material to enable your device function or a very tight feature control during an etch process, you must plan to devote additional resources to process development during early prototyping.

In this chapter, we introduce the following strategies for success:

- Make a MEMS process flow lower risk by composing it from building blocks of common process modules.
- Design for realistic fabrication conditions and volume production in order to establish process flow success early on.
- Best practices for photomask layout and lithography: well-organized, forward thinking mask data will save many headaches later on.

This is just the tip of the MEMS process development iceberg. We refer you to many other excellent textbooks which capture the large body of information on the many details of MEMS fabrication processes (See Table 11.2).

© The Author(s), under exclusive license to Springer Nature Switzerland AG 2021
A. M. Fitzgerald et al., *MEMS Product Development*, Microsystems and Nanosystems, https://doi.org/10.1007/978-3-030-61709-7_11

Assembling a MEMS Process Flow

To begin, this section defines some terminology to help you understand the unique challenges with MEMS fabrication. Examine the marketing materials for any CMOS foundry and you will see lists of available, standardized *process flows* at different lithography nodes (e.g., 0.18μm CMOS). You select a CMOS foundry based on a given CMOS process that meets your needs and then provide a mask file, containing your design, which has been checked against the foundry's specific design rules for that process. For readers less familiar with this technology, imagine a bakery offering a cake recipe that can be baked in the form of any alphabet letter. You tell them which letter (the layout) you want and they will bake a cake with their proven recipe (process flow).

In contrast, marketing materials for MEMS foundries contain lists of available *process steps* (e.g., DRIE, oxidation, electroplating). When working with a MEMS foundry, you provide both a 2D mask file and a process flow to execute the design, composed of the process steps offered by the foundry. Extending our previous example, here the bakery lists recipe steps they can execute (e.g., sifting flour, whipping eggs whites, and greasing a cake pan) that can be combined to form a recipe. You supply the shape of the letter (the layout) and the desired recipe steps organized into a unique recipe to bake the cake (process flow). The lack of standardized MEMS process flows, analogous to what CMOS foundries offer, is a long-standing reality in MEMS fabrication that was most concisely summarized by Jean-Christophe Eloy of Yole Développment: "one product, one process, one package" (see Chap. 1).

Instead of set process flows, MEMS fabrication employs process modules, which are then organized to create unique process flows which produce unique products. Think of modules as LEGO® blocks which can be reconfigured in a large number of combinations in order to create unique process flows. Table 11.1 illustrates the hierarchy of a MEMS process flow based on MEMS process modules which are, in turn, based on individual process steps and their recipes.

We outline some helpful guidelines that can be implemented at the lower levels of the process hierarchy in order to help reduce risk and costs. (Chap. 15 outlines options for mitigating risk at the process flow level.)

Table 11.1 Hierarchy in a MEMS process flow

Level number	Level name	Description
1	Process flow	Collection of process modules that accompany the mask layout to make a MEMS device, often named for the final device type or design iteration (e.g., pressure sensor process flow)
2	Process module	Collection of process steps to achieve a desired feature (e.g., photolithography)
3	Process step	Steps or "wafer moves" that are executed in a particular order to complete a module (e.g., spin coat photoresist)
4	Process recipe	Details to execute the process step on a wafer (e.g., type of photoresist, dispense time and volume, spin speed for desired thickness)

Any change made at any process hierarchy level creates a new and unique process flow. In MEMS processing, due to the large number of variables at play in each process recipe, and interactions with preceding and subsequent process steps, there is no such thing as simple tweak or change to a process flow. Sometimes seemingly minor changes, such as slight change in a feature dimension, can have dramatic effect on device performance. Significant process flow changes, such as changing or introducing a new process module, adding new materials, or using a different process tool, can have large impacts that will send you backwards to redevelop the process flow.

On a more positive note, sometimes small modifications to a robust process flow can enable a family of related products. Once you have a baseline process flow that yields well, your device may be differentiated through minor process tweaks into a multitude of products. For example, a family of pressure sensor products could be created by fabricating different membrane thicknesses and/or areas. In order to attain that capability, you would first need a solid understanding of the entire process flow operating window, to know where you can safely make small changes, and at what point a process change could cause problems.

Common MEMS Process Modules

This book assumes the reader already has a basic understanding of MEMS process methods. Our goal here is not to delve into every detail of MEMS processing but to raise awareness of these higher-level topics in process development planning. If any of the process technologies listed in this chapter are unfamiliar, we recommend the resources listed in Table 11.2. (A list of resources related to MEMS design can be found in Chap. 10.)

Table 11.2 Recommended references for detailed technical information on MEMS process integration and fabrication methods

Topic	Title	Reference
Overview of MEMS fabrication processes	An introduction to microelectromechanical systems engineering	[1]
Describes all standard MEMS process steps	Fundamentals of microfabrication and nanotechnology	[2]
IC fabrication processes, which formed the basis for early MEMS processes	Silicon processing for the VLSI era	[3]
Key MEMS process details	MEMS materials and process reference handbook	[4]
Key MEMS process details	Handbook of silicon based MEMS materials and technologies	[5]
Information on a step that enables many MEMS devices, wafer-level packaging, and requires very specific process conditions	Wafer bonding	[6]
Managing dimensional and material property variations which occur in micro- and nano-fabrication	Process variations in microsystems manufacturing	[7]

The typical process modules that are used in MEMS fabrication are shown in Table 11.3. For each type of MEMS device, these modules are combined in unique and different ways in order to create the process flow.

The number of possible combinations is endless. Even for a specific type of MEMS device, it is possible for completely different process flows to create two devices having similar function and performance. Figure 11.1 shows cross-section SEM images of two different commercially available dual-axis MEMS accelerometers. Each of these sensors were made by different companies, by two completely different process flows and mask designs, yet both process flows created accelerometers having very similar performance.

Table 11.3 Typical MEMS process modules

Process module	Description
Oxidation	Growth of a thin silicon oxide layer film by chemical means
Deposition	Deposition of a thin layer by physical or chemical vapor or mechanical means
Lithography	Transfer of the pattern from a photomask or maskless writer to a wafer using a photo-definable polymer, known as photoresist
Etch	Selective removal of material, usually defined by the pattern transferred during lithography
Sacrificial etch	Removal of material, usually without a photo-defined mask and determined by time or material selectivity
Bond	Temporary or permanent contact between two wafers
Dicing	Separation of individual die after wafer-level process is complete

Bulk Micromachined Accelerometer   Surface Micromachined Accelerometer

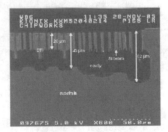
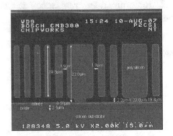

Key process steps:
DRIE into Si wafer
XeF2 etch

Key process steps:
Oxide and nitride deposition
Thick polySi deposition or growth
DRIE of polySi

Fig. 11.1 Accelerometers having similar performance, yet fabricated with two completely different process flows (Reproduced with permission from TechInsights, formerly known as Chipworks, 1891 Robertson Road, Suite 500, Ottawa, ON K2H 5B7 Canada)

Table 11.4 Factors to consider when integrating common MEMS process modules in order to minimize process risk and complexity

Process module	Integration factors to consider
Oxidation	Thickness and uniformity tolerances, film stress, thermal budget
Deposition	Thickness and uniformity tolerances, film stress, thermal budget, topography
Lithography	Contact vs. stepper, feature resolution, reticle field size, alignment strategy, backside alignment, topography
Etch	Mask and material selectivity, aspect ratio, topography, etch stop control
Sacrificial etch	Etchant selectivity, cavity aspect ratio, topography
Bond	Yield trade-offs; top to bottom alignment strategy; thermal budget, depending on method
Dicing	Size of die, fragile released structures, materials, space for test structures, keep out zones

This broad flexibility in design both at the layout and at the process flow levels has created a wide variety of novel MEMS devices. It is also been used as a strategy to navigate through and around a crowded patent landscape. But getting too carried away with creative process flows can also increase development complexity and therefore costs. The guidelines outlined in this chapter can help you corral all the available information and keep the overall process complexity to a minimum. Important factors to consider when integrating MEMS process modules are outlined in Table 11.4.

If the factors listed in Table 11.4 are unfamiliar, we recommend referencing the texts in Table 11.2.

Designing for Realistic Fabrication Conditions in Volume Production

Planning for manufacturability of your MEMS device starts early in process development. If your device is designed with realistic process and manufacturing tolerances, then your development path to volume manufacturing will be shorter and less expensive. If you have created a device design that is ignorant of or deliberately pushes tolerances, we highly recommend you take a staged approach to development: starting with easily achievable process tolerances and then tightening them only after you have established the proof of concept and produced some sample die to enable continued product and system development. If you aim for tight process tolerances and device specifications from the start of your development, you may end up with no usable prototypes, even after several prototype runs! You also do not want to get too far into device development only to realize that MEMS foundries will not be able to consistently or cost effectively replicate your design because it relies on conditions which are difficult to achieve.

A typical area where understanding realistic process conditions is essential is in the design of suspended structure devices, such as a membrane or resonant beams. The performance of these devices is highly sensitive to the stress of any thin films deposited onto the suspended structure. Thin films that are more tensile than expected can significantly shift the structure's resonant frequency, and those that are overly compressive could buckle the structure. You must be aware of typical stress values for the materials you intend to deposit and factor that into your structural design.

For example, if depositing a PECVD nitride film, the structural design must account for the film stress achievable by the deposition tool. If the design were to depend on a film having stress of 75 ± 10 MPa, but the tool's two-sigma process window is 100 ± 25 MPa, then your design will be doomed to low yield because the tool will only occasionally be able to hit the lower end of its process window (Fig. 11.2).

If you have inherited an academic process flow, the mismatch between the design requirement and the process window is especially important to evaluate. In academia, a low-yielding process can still be considered a significant success if the device developed is truly novel and worthy of scientific or engineering recognition in peer reviewed academic journals. In the commercial world, it is a recipe for missed cost targets and limited market opportunity.

You must also balance trade-offs in process choices based on the end use market and the potential for process flow changes. For example, once FDA-approved, the process to fabricate a MEMS medical device could not be changed without regulatory paperwork burden and risk. Understanding these constraints and trade-offs for your device can help you to determine where you might need to focus your attention: for example, reevaluating the design, using processing risk mitigation strategies outlined in Chap. 15, or developing in-house tools or processes to achieve desired features.

In general, it is much easier, faster, and cheaper to design a MEMS process flow within the capabilities of an existing process module or step versus trying to force it to conform to requirements that are at the edge of a process window.

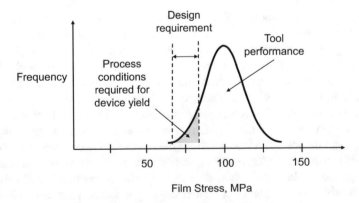

Fig. 11.2 Process window and design requirements contribute to yield

Using "Secret Sauce" Processes

If developing in-house tools or proprietary processes, sometimes casually described as "secret sauce," the process's portability, foundry availability, including second sources, cost, and reliability, are just some of the factors that should be considered. If your device's "secret sauce" happens to be controlled by a third party (for example, vacuum getters or other specialty materials), this adds another layer of complexity to your process, and ultimately to your business. If a supplier develops a "secret sauce" for you, or sells you their proprietary materials, you will be dependent upon their sole source supply. Using secret sauce should be limited to those processes that would truly differentiate your product in its performance, application, size, and that, very importantly, whose cost can be supported by the market (see Chaps. 4 and 5).

Ideally, it should be the device design that makes your device truly special, rather than relying on unusual or special process recipes. If you can avoid incurring costs and long timelines of developing your own special fab processes, do so! Table 11.5 outlines the pros and cons of using secret sauce processes to help you analyze some typical trade-offs.

If you are unable to develop a process that is well within the MEMS process technologies common to many foundries, you may want to consider shifting early development of your device to a carefully selected production foundry. Processes could then be developed on the specialized tools that would be needed in volume production, with the process development knowledge and history retained by the foundry team. This situation occurs, for example, with piezoelectric materials, in which the performance of the piezoelectric is very specific to the deposition tool and process recipes. If your design might need a specific formulation of PZT, for example, that Foundry A excels at, you should consider developing your product there. Doing so may impede the portability of your design to other foundries or possible second source suppliers; however, getting the material your design needs, quickly, may be worth that trade-off. (Starting a development at a production foundry will

Table 11.5 Pros and cons of using common versus "secret sauce" MEMS processes

	Pros	Cons
Common MEMS processes	• Portability between foundries • Faster development time • Less expensive processing • More expertise available	• More easily copied • Can limit technology performance • More competitors in your space • Product differentiated through mask design only
Secret sauce processes	• Barrier to entry (no one else can do it) • Differentiating or unique performance • Might open other business opportunities such as licensing, etc. • Trade secret protection	• Lack of second source or difficult to replicate elsewhere • Requires experts • Cost of process development • Potential regulatory issues with unusual materials (RoHS, conflict minerals, etc.)

require you to have your business case well established early on, as explained in Chap. 17.)

You may also not yet know at which process tolerances your device design would still function, so you might start at the edge of the "process window," and through iteration move the process towards more common and easily achieved specifications. Understanding how close your device is to the edge of a process window, and making an effort to fix that during development, will help to avoid unpleasant surprises of low yield when moving to volume production.

We recommend against giving into the temptation to use any non-scalable or manual processes in the prototype stage as a short cut. For example, using manually applied Kapton tape to mask a key feature during a plasma etch process, or a hand scribe and break to remove a part from a silicon wafer. This type of process hack may be tolerable at the proof-of-concept stage, but once in the advanced prototypes stage, where customer evaluation, system development, testing, and qualification are potentially happening, take the time to properly develop a process that can be scaled.

Creating a Process Flow Runsheet

During prototype development, the MEMS process runsheet, which documents the execution of the entire process flow, should not only contain the necessary information to run the process (number of wafers, process tool, process recipe, recommended recipe input variables, etc.) but also contain instructions regarding material specifications, and inspection instructions, such as which features to measure and when (e.g., lateral dimensions, films thicknesses, etch depths, sidewall angles, etc.).

Referring to the MEMS process flow hierarchy defined in Table 11.1, each line in the runsheet should correspond to a process step (Level 3) and contain the process recipe (Level 4) along with any additional relevant information.

It should also provide space to capture the results, observations, and general experience for each process step. Examples of important process notes to capture during development include:

- Color of films after deposition or during an etch process.
- Layout features that etch faster than others, such as due to etch non-uniformity.
- Measured etch rates at multiple points on a wafer.
- References to images of processed wafers.

It is also helpful if the runsheet format allows documentation of the processes used for risk mitigation such as look ahead wafers, batch splits, and staging wafers (See Chap. 15). The exact format style of a runsheet is not critical; however, it should make it easy to collect information in an organized way during what can be a busy time. A basic example is shown in Table 11.6.

Development runsheets can become messy due to wafers splits, abandoned wafers, hastily written data, etc. At the end of a wafer process run, review each

Table 11.6 Example runsheet format. (Note: rows are intentionally left blank)

Step #	Description	Tool	Recipe and Input Variables	Tolerances Pass/Fail Criteria	Inspection Details and Processing Notes	Wafer IDs
1						
2						

runsheet for missing data, add clarifying notes, and identify what information should be passed on to future process runs, and eventually to the production foundry. This may also include data from destructive inspection after the wafer process has been completed, such as SEM cross-section analysis of DRIE sidewall angles. You should also review any device performance test data to establish links to process tolerances.

The vision that comes from process hindsight, coupled with device test data, can be a powerful tool for informing design iterations and future runs. Test data may, for example, show that a 10% over etch is acceptable for the device performance, but 15% would be too much. Taking the time to do this detailed documentation after each process run will not only capture learning during prototype iterations but also help when you are ready to move to foundry production.

Leveraging Foundry Process Platforms

Some MEMS foundries do offer standard process flows for a limited set of MEMS devices, including pressure sensors, gas sensors, gyroscopes, and accelerometers. If you have the ability to utilize these platform processes in order to fabricate your product, you will enjoy a faster time to market and significantly reduced development costs by leveraging production-proven foundry process modules, as illustrated in Fig. 11.3. In most cases, the foundries may be willing to consider small modifications to their platform process to allow you to customize it to your desired specifications. If you find a process platform that addresses a significant portion of the capability you need, it is worth talking to the foundry about your options (Also see Chap. 7 for more information on leveraging third-party IP).

Best Practices for Lithography and Photomask Layout

In the photomask layout, there is the most design control and the most freedom, and therefore a good chance for a beginner MEMS designer to get into trouble. It is always possible to draw anything "on paper." Designing within the capabilities of a process step will be embodied in the mask layout. It is therefore really helpful to understand a few key strategies about lithography and how to set up your photomask layout file.

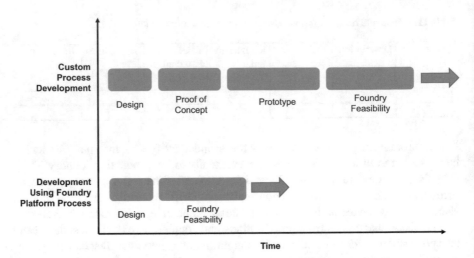

Fig. 11.3 Using a foundry platform process enables you to skip the proof-of-concept and advanced prototypes stages and save significant time (>1 year). This huge time savings may be well worth the trade-off of being dependent on that foundry for future supply

This section assumes a basic understanding of common MEMS lithography methods. We only summarize the physical differences between methods and instead focus on trade-offs between different options and the differences between what is used in prototyping versus in production.

Lithography Methods and Parameters

The two types of photolithography methods that are most commonly used in MEMS R&D fabrication are contact and stepper lithography[1] . With the exception of large chip (>22 × 22mm) MEMS, such as microfluidics, many foundries prefer stepper lithography for production. The yield is usually higher because the mask pattern is projected onto the wafer and therefore there is no contact between the mask and the wafer, reducing the risk of damaged photoresist, mask wear and tear and debris contamination. In addition, stepper lithography also offers more precise layer-to-layer registration and feature size tolerances than contact lithography.

In the proof-of-concept and advanced prototyping phase, both contact and stepper lithography are regularly used. Contact lithography provides a cost-effective way for multiple chip design variations to be placed on the wafer. Whereas with stepper lithography, due to projection optics, the photomask field is limited for

---

[1] Other forms of lithography, such as E-beam lithography and nanoimprint lithography, are sometimes used in MEMS, but less common. Please see references in Table 11.2 for additional information on alternative lithography methods.

design variations because the photomask pattern may be 4–10 times larger than the pattern will be on the wafer.[2]

Beyond the production considerations already mentioned, the lithography method is usually chosen based on the required feature sizes. Contact lithography can reliably expose and resolve features down to 4μm, and achieve 2μm with attention to exposure conditions. Trying to resolve anything less by contact lithography is pushing the process window and will therefore require processing a large number of wafers in the hopes to get a few good ones.

Stepper lithography can easily achieve sub-micron linewidths, the exact minimum feature resolution being dependent on the tool technology generation. In general, 0.5-μm linewidths can be achieved by stepper tools that are commonly used in MEMS production foundries. If you need smaller features, you will likely need to select your development and foundry partners from a limited pool based on the availability of the specific lithography capability.

Photoresist

In addition to the lithography method, the achievable feature size is also determined by the type of photoresist, the thickness of the photoresist, as well as the topography of the wafer surface. Thicker photoresist and larger topography limit the minimum definable dimension. A general rule of thumb is to have no more than 1:1 ratio between the thickness of the photoresist and the width of the line or space in the photoresist, although slightly larger aspect ratios may be possible. If your design's minimum feature size is 2μm, your photoresist should be no thicker than 2μm. If your resist is 10μm thick, defining 2μm features may not be possible, and if it is, will most likely require special photoresist and exposure parameters, which means it will be more expensive to process.

With regard to doing lithography over topography on a wafer, a common situation in MEMS, you will have difficulty achieving good resolution of your feature if you are exposing a pattern on two co-planar surfaces that are located more than a few microns apart in the vertical direction. For example, it is possible to adjust the focal plane in stepper lithography, but that would require two exposure passes with different exposure settings.

The minimum thickness of the photoresist is also determined by how much is needed to mask the desired material. For etching, the thickness depends on the selectivity of the etch process between photoresist and the material being etched. For instance, this selectivity can range between 1:1 and 100:1 (the photoresist etch

---

[2] In contact lithography, the photomask has the entire wafer layout. In stepper lithography, the photomask has a portion of the layout that is repeated across the wafer during exposure of the photoresist. The term photomask is used in this chapter to refer to both types of lithography and to also differentiate between it and the photoresist mask on the wafer. Photomasks are also called reticles, in the case of stepper lithography.

rate is the same, or up to 100 times slower, than the material's etch rate, respectively). Different photoresists have different resistance to etching, and some tools have a passivation step that helps maintain the photoresist during a plasma etch. Because the resist selectivity is process and tool dependent, specifics regarding photoresist choice is usually left up to the development fab or foundry to select. In the case of liftoff processing, a general rule of thumb is that the photoresist should be 2–4 times the thickness of the material being deposited, depending on whether the resist has an overhang or re-entrant sidewall to prevent sidewall deposition.

By now you should be getting the gist of the following: the more relaxed the constraints are for the aspect ratio of the photoresist, the more robust and flexible your lithography process is and therefore the less expensive it will be to develop and to process. Unless it specifically provides your device its novel capabilities or is the functional keystone, your MEMS development should not require the cutting edge of lithography technology.

Fortunately, there are many photoresist options to choose from for your device: negative vs. positive photoresist; photoresists with different light bandwidth sensitivity and selectivity to different etch processes; photoresists with different viscosities for different thicknesses and topography coverage, etc. More details regarding photoresist types can be found in the resources referenced in Table 11.2.

Alignment Strategy and Marks

With so much of the design effort focused on the actual MEMS device, the alignment strategy for each photomask layer can sometimes be an afterthought. It can often, however, take a significant amount of time to ensure each photomask will be correctly aligned to patterns from previous process steps. In R&D environments, alignments between mask levels is often done manually in contact lithography, where the person processing the wafers aligns the mask by viewing marks through a microscope or on a video screen. In this case, alignment marks must be created individually for each layer, taking into account the mask polarity and the requirement to see through the mask to the pattern on the wafer.

When using a stepper exposure, the mask is first aligned to the tool itself using features outside of the device area. These alignment features are unique to each exposure tool and are provided by the tool manufacturer. The tool is then aligned to the wafer again using a unique pattern provided by the tool manufacturer.

No matter the lithography method used, it is critical for either the human or the machine to be able to see the alignment mark pattern on the wafer. Part of developing an alignment strategy is making sure the subsequent process steps will not cause alignment marks to be obscured or removed, such as by a deposition of an opaque metal film or a polishing process, respectively.

Backside alignment, a process that is often used in MEMS where structures are built by stacking wafers, using engineered substrates or wafer-level packaging, also

Table 11.7 Example of an alignment mark check table (Note: rows intentionally left blank)

Layer (GDS #)	Data: digitized dark or clear	Material removed (etch) or deposited (liftoff)	Aligns to layer (GDS#)	Acceptable alignment tolerance	Clear view of previous layers' marks?	Clear view of target mark through photomask?
1						
2						
3						

adds another layer of complexity with alignment features often needing to be located in very specific positions on the wafer.

Because of all of the factors at play, developing an alignment strategy involves some effort to check the process flow, mask data, and photomask specifications in detail to make sure that each alignment step will proceed as planned. Without careful attention to alignment strategy and marks, you may find yourself needing to buy a replacement mask mid-process or, worse, redoing an entire wafer process flow.

Using a table, such as shown in Table 11.7, can help to organize the contributing data and to reduce the chance of mistakes. It is helpful to create such a table before you even start your device mask layout. It is also helpful when working with multiple contributors to a mask layout design. Finally, providing your mask data to colleagues to double check is a critical review step that we highly recommend.

Setting up the Photomask Layout File

When drawing the MEMS mask data in your layout/CAD program of choice (commonly used layout tools include: Cadence, L-Edit, AutoCAD, SolidWorks, and KLayout), take the mindset that you will likely need to update it again and again. This means adopting behaviors such as:

- Automating feature generation, especially for repetitive patterns, when possible (if you do not have the skills, the software vendors are often happy to write a macro layout program for a fee).
- Using instances and hierarchies, which almost all mask layout software employs (e.g., in Mentor Graphics' L-Edit, they are called cells).
- Naming your layout components/cells and overall layout files using clear, easily understood terminology (e.g., 4_μm_comb_finger vs. comb_finger_A for a component that makes up an electrostatic comb drive of which there are three versions with varying finger widths) so you can easily keep track of inevitable changes[3].

---

[3]This example would require you to update the cell name if the comb finger width is changed. If the dimensions are not set, you could also use small, medium or large for your three comb widths. Either of these options are easier to remember and quickly comprehend than "A, B, or C".

For MEMS, the layout dimensional unit is usually microns, with the layout programs providing an additional three decimal places for nanometer resolution (e.g., 1.234μm). Even for stepper lithography, where the mask dimension will be 4–10 times larger than the final dimension on the wafer (depends on the projection ratio of the specific tool), it is helpful to draw the mask data using the final dimensions desired on the wafer. It is standard practice for photomask vendors to scale the mask data to the appropriate size depending on the lithography tool that will be used.

Whether using contact or stepper lithography, the photoresist mask, and therefore the device, will not have single digit nanometer resolution. The final resolution achieved will depend on the lithography method, the type of photomask and how is it made[4], the photoresist and subsequent processing. Every photomask layout software will have its own dimensional setup, so first check if there is a manufacturing grid that should be assigned to the mask data file.

For example, in L-Edit, if the manufacturing grid is set to 0.1μm, all exported GDSII files, the file format preferred by most photomask vendors and foundries, will have the data set to a 0.1-μm grid. Any vertices that are off this grid will be moved to the closest tenth of a micron. When inspecting the exported GDSII file that will be provided to the photomask vendor, you must check that all the features have been rendered as expected. If this data grid is set too large, for instance at 1μm for a 5-μm circular feature, you will probably find that your well-defined curve looks more like a polygon (example shown in Fig. 11.4).

To prevent any undesirable geometry shifts in the exported data and final photomask, when drawing your layout, it is helpful to draw your features aligned to this defined grid. Drawing to the grid also helps prevent mistakes, makes checking the layout easier and building your file hierarchy easier. If a 10-μm square is drawn with its base vertex at location (0,0), it is much easier to do calculations and work with than if it was drawn at location (0.035, 0.007)! Investing attention to these small, seemingly insignificant details will pay dividends because you will very likely find yourself keeping track of a lot of coordinate numbers as you build your photomask layout.

Another helpful action to take before you start drawing the mask layout is to think about how the layout of your device should be structured using the program's organizational hierarchy. Using L-Edit's terminology as an example, it is helpful to create a wafer-level cell that will either contain the final wafer-level layout for a

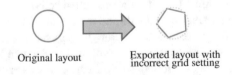

Original layout   Exported layout with
                         incorrect grid setting

Fig. 11.4 Photomask data file manufacturing grid affects how curves will be rendered in exported GDSII mask files and on the photomask

---

[4]In addition, the references listed in Table 11.2, many photomask vendors provide simple primers how photomasks are made, for example, Compugraphics has an education center [8].

contact lithography mask reticle, or to visualize how the final wafer layout will look after exposure with stepper lithography. The corresponding wafer layout for stepper lithography would use a cell that contains the final photomask reticle step and repeat layout. Creating and using a wafer-level template cell that may be reused among your design team is even better. This cell can contain valuable information such as the wafer outline, where alignment marks should be placed for different processes or tools, and perimeter keep out zones. Sub-cells also included in this template could include commonly used alignment features or test structures.

When it comes to organizing the file structure for your mask layout, it will be dependent on the type of MEMS device. For example, a MEMS device that has a large number of metal bondpads that connect to a doped region in the silicon substrate through an oxide layer, it may be helpful to create a cell that contains the pattern of a contact etch through the oxide and of a metal bond pad. Inserting and aligning a cell instance will take much less time than drawing the feature at every location, especially after the twentieth bondpad. But the power of using these hierarchical cells to create instances goes further when it is determined that the bondpad size needs to be decreased or increased. If you have used the cell hierarchy, this change is made once and then automatically propagated through the rest of the instances in the entire layout. The cell hierarchy is also useful when dealing with layout variations. If you have a device design that uses cantilevers of varying widths, you can create individual cells with each of these cantilevers and then insert each as a sub-cell into a top-level cell as needed without drawing each cantilever over and over again, as shown in Fig. 11.5.

With this great power, however, comes great responsibility; if a change is made in an instanced cell, you will need to check the overall layout for unintended consequences of that change (e.g., the larger bond pad now interferes with a feature on another process layer which might prevent it from being properly etched). We usually find that the benefits outweigh the risks when working with hierarchical instances.

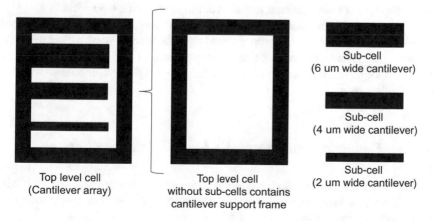

Top level cell
(Cantilever array)

Top level cell
without sub-cells contains
cantilever support frame

Sub-cell
(6 um wide cantilever)

Sub-cell
(4 um wide cantilever)

Sub-cell
(2 um wide cantilever)

Fig. 11.5 Example layout with cell hierarchy for design variants (array with cantilevers of varying width)

Design Variants

Design variants are a key part of the proof-of-concept and advanced prototypes stages and can be a powerful tool for exploring the physics of a new device. However, you should exert some restraint and keep the number of design variants to a reasonable number. If you feel tempted to create more than 5 design variants on your photomask layout, you should consider dealing with your feelings of uncertainty by instead doing more simulation work on the device design. A large number of design variants on a photomask can create confusion, especially if they are not properly labeled and tracked.

The appropriate number of design variants may be determined by analysis of the design space (such as by using the method of design of experiments [9]), a process risk analysis (see Chap. 15), or, simply, the number of devices you can actually test. Promise that if you draw it, you are going to later test it and document the results! Otherwise, you are just wasting your time on useless masks designs and unnecessarily consuming valuable silicon real estate.

Design variants must be chosen thoughtfully, because they can affect the outcome and success of a process step. For example, design variants with varying feature line and space widths can put your entire wafer at risk, if in trying to get one set of designs to finish etching, you overetch all the others. It helps to have a very clear understanding of which variants are the priority designs and which designs might be there in order to explore the limits of a process (e.g., how large of an aspect ratio can you achieve) or as a hedge to yield some working devices, if a process window is not well understood.

Lastly, when organizing the overall wafer layout for contact lithography, it is helpful to distribute the design variants across the wafer instead of clustering them all together. If you were to have yield loss on one section of a wafer, you would not have lost all die of a design variant. If a process has a cross-wafer non-uniformity, distributed design variants can help characterize the non-uniformity or even inform you on how best to bias the next mask layout in order to improve yield.

Once you are in the foundry production development path, the design variants should be used exclusively to determine and tune the process window, not to further optimize the device design. With few exceptions, you will not be able to enter foundry pilot production until you have selected a single design around which to optimize the process.

Testing, Packaging, and Assembly Layout Considerations

In addition to the MEMS device and the alignment marks, it is important to also consider test, package, and assembly requirements and how they might affect the device and wafer mask layout.

Test structures include structures for the purposes of process control monitoring (PCM), device performance characterization, and troubleshooting. These structures can be placed in dicing lanes, adjacent to alignment marks, or even in a "blank" device die location. Understanding wafer space constraints from the start can prevent you from having to redraw the structures once you are determining the final wafer layout.

Similar to design variant placement, it is best to have the test structures distributed across the wafer, in case of partial yield loss or wafer processing non-uniformity; the bare minimum is one set per wafer quadrant. Also consider how the MEMS itself will be tested (e.g., manual probe, wafer probe card, packaged). The layout for bondpads can be very different for manual (larger, non-uniform pitch, etc.) vs. automated probe testing (uniform layout for all die). See Chap. 12 for more information. If the die will require manual handling in the early development phases, consider creating a larger area die at first while planning for future die shrink.

With regard to packaging and assembly, in most cases, the wafer will first need to be singulated into individual die. It is critical that you know the dicing method that will be used to singulate your wafer before you design your wafer-level layout. The method will determine the spacing, or streets, required between die. If not planned appropriately, your die could end up smaller or larger than desired, or even worse, you could lose functional components on the die. For mask layout purposes, in traditional saw dicing, dicing blades can take 18μm to 100μm of material from the wafer (known as the dicing or saw kerf). The blade thickness is selected based on the wafer material type and thickness. In plasma dicing, the dicing kerf is determined by the allowable aspect ratio of the etch process. Stealth dicing is a zero-kerf process; however, a dicing lane free of materials that block IR laser light is required. (More information regarding dicing can be found in Chap. 12.)

If you plan on using a commercial off-the-shelf (COTS) package for your device, we strongly recommended that you specify and select it before the device die layout is finalized. This can save you the headache of later finding out that your die is 0.1 mm too large for the otherwise perfect package or if the same device area in a different form factor (rectangular vs. square) could have saved you money on packages. For custom package development, it is ideally a co-development path with the MEMS device that will require communication throughout the design process.

Planning for Future Die Shrink

Do not be afraid to make early proof-of-concepts die larger than they will need to be in full-scale production. The form factor for the first prototype does not have to be the same as the final product and this can make everyone's life easier: the process engineer, the test engineer, the packaging engineer, etc. Extra spacing for handing wafers, reducing etch load, providing larger contact probe pads, being able to use standard pick and place tooling, etc. will allow you to focus on getting the device operational as quickly as possible without making anything harder than it needs to be in the early stages. If you need to generate more device die, plan to process more wafers rather than cramming more die onto a wafer layout. Running more wafers has an added benefit of providing more data for future processing runs and helps with risk mitigation on multiple levels (see Chap. 15).

It is still important to address the long-term wafer processing and assembly plan, by outlining how the device size can be reduced over time, paying particular attention to items that could require modification to the process flow and packaging method. How quickly you need to show functionality in your device's final form factor depends on your product requirements and application and how disruptive the technology might be. If you are designing a pressure sensor for a 1F catheter, then the size is the most critical feature of the device and should be met in the first prototypes. If you are designing a micro-mirror array, you can start with a smaller array of mirrors to minimize process complexity and risk, and then increase the number of mirrors in subsequent process runs as more information and understanding is gained.

Finalizing a Mask Layout

As previously discussed, your (2D) mask layout is part of a (3D) process flow. MEMS engineers develop the ability to visualize the 2D layout as a 3D structure in their head, as well as the cross sections that are often used to illustrate a process flow. However, even experienced MEMS engineers can benefit from using a software program that enables 3D visualization of the final device.

In addition to helping colleagues, executives, and investors, 3D process visualization software can help to identify potential mask or process issues before fabrication begins. For example, 3D rendering of a process could reveal that misalignment between features on different layers could create an unintended access point for an isotropic release chemical. You can think of 3D process visualizations as a type of design check that can be employed to virtually investigate and validate new process flows.

When you are ready to create a manufacturable version of your mask file, a.k.a. "tape-out," no matter which mask layout software was used, you will likely need to export your data into the GDSII file format. It is the data format of choice for most photomask vendors and foundries. The mask layout engineer must always check that the layout data was exported to GDSII format correctly. We recommend using a different layout viewer to perform this check than the software that was used to create and export the layout in the first place, because errors in the data can sometime be obscured by the original software's import functions.

The GDSII layout should be reviewed and checked by someone on your team who is familiar with the device design and process and who can bring fresh eyes to inspect the mask layout. It is helpful to provide your mask-checker with the following information:

- Original layout file (native to the program used to create the layout).
- Exported GDSII file ready for the mask vendor or foundry.
- Process flow outline (cross section diagrams or the 3D visualization can be extremely helpful too).

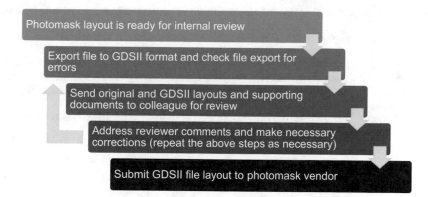

Fig. 11.6 Photomask layout "tape-out" review flow chart

- Mask checklist (number of design variants, key dimensions, exceptions, etc.)
- Table with order data to be provided to the photomask vendor or foundry (polarity of data, location of critical dimensions, mask resolution).

This process is summarized in Fig. 11.6. Gathering this information for your colleague provides the added benefit of creating essential documentation of your layout, as outlined in Chap. 16. Once the mask check has been successfully completed, you are ready to submit your layout to the photomask vendor, and look forward to wafer processing.

Summary

In today's MEMS technology landscape, designing a process from textbook knowledge is quite different from designing a MEMS process to execute in real processing tools. This chapter only scratches the surface of the information you will need to successfully perform a process integration and mask layout. For best results, use standard MEMS process modules whenever possible and design your device for realistic process conditions and dimensional tolerances. If you have inherited a prototype process flow, reevaluate it with these manufacturing factors in mind. It is usually worth the effort to redesign if too near the edge of a process window.

References

1. Maluf N, Williams K (2004) An introduction to microelectromechanical systems engineering, 2nd edn. Artech House, Boston
2. Madou M (2011) Fundamentals of microfabrication and nanotechnology, 3rd edn. CRC Press, Boca Raton
3. Wolf S, Tauber RN (2000) Silicon processing for the VLSI era, Vol. 1: process technology, 2nd edn. Lattice Press, Sunset Beach

4. Ghodssi R, Lin P (eds) (2011) MEMS materials and process reference handbook. Springer, New York
5. Tilli M, Paulasto-Korchel M, Petzold M, Theuss H, Motooka T, Lindroos V (eds) (2020) Handbook of silicon based MEMS materials and technologies. Elsevier, Amsterdam
6. Alexe M, Gösele U (2004) Wafer bonding, applications and technology. Springer, Heidelberg
7. Huff M (2020) Process variations in microsystems manufacturing. Springer Natire, Switzerland
8. Compugraphics (2017) Education center. https://www.compugraphics-photomasks.com/education-centre. Accessed 5 Aug 2020
9. Box G, Hunter J, Hunter S (2005) Statistics for experimenters: design, innovation, and discovery, 2nd edn. John Wiley & Sons, Hoboken, NJ

Chapter 12
Design for Back-end-of-Line Processes

The back-end-of-line (BEOL) manufacturing processes, also called the "back end," are easily overlooked in the early stages of MEMS design[1]. However, in order for a MEMS device to be manufactured into a product, the die must be designed with the back end processes in mind. Too often, the requirements for these processes are not fully considered until after prototype wafers have been completed, at which point it is realized that they do not meet the BEOL requirements, thus requiring redesign. To understand how to best design MEMS products for the back end, we give an overview of typical MEMS back end processes, while discussing ways to incorporate back-end requirements in product design.

After the wafers are completed by the foundry, the typical back end steps are as follows:

- Wafer sort.
- Wafer thinning (optional).
- Singulation.
- Packaging and assembly.
- Package test.

We discuss the first four of these steps in this chapter. The last step, package test, is discussed in Chap. 20.

---

[1] In CMOS processing, the terms "back-end-of-line" and "back end" refer to two distinct sets of processes, unlike in MEMS vernacular, where they may be used interchangeably. In CMOS, "back-end-of-line" refers to the latter stage of wafer processing during which metal interconnects are fabricated, and "back end" refers to the post-foundry processes, common also to MEMS, which are described in this chapter.

© The Author(s), under exclusive license to Springer Nature Switzerland AG 2021   149
A. M. Fitzgerald et al., *MEMS Product Development*, Microsystems
and Nanosystems, https://doi.org/10.1007/978-3-030-61709-7_12

Wafer Sort

The first BEOL step is wafer sort, also called wafer probing. The wafer is held by a vacuum chuck, and a set of probes makes electrical contact to the pads on the die, which are connected to instruments that can test the die. See Fig. 12.1.

The purpose of wafer probing is twofold. One is to test and confirm that the wafer as a whole is a good wafer, was correctly processed, and meets requirements. The second is to identify which die on the wafer meet pass/fail requirements. Wafers that fail are removed from the manufacturing stream and undergo failure analysis. The results are then used to improve the wafer fabrication processes. The die that fail are either marked with physical ink or, more commonly, mapped electronically to a wafer map. After wafer singulation, the bad die are discarded or undergo failure analysis.

The types of tests that can be done at wafer sort are usually only electrical. The electrical tests include checks of opens, shorts, resistance, capacitance, and inductance. Higher-level tests are possible, such as capacitance-voltage (C-V) curves and resonant frequency responses. Specialized MEMS wafer probers can also apply other kinds of stimuli, such as pressure, sound, and light.

The wafer sort data is rich, and there are a number of useful insights that can be gained by statistical analysis of the data. Since the number of die on a wafer is typically on the order of thousands or tens of thousands, the statistical power of the probe data is high. In addition to averages, standard deviations, and other statistical values, the histograms of the tested values should be examined. The distributions are typically Gaussian in nature. Multimodal peaks in the distribution can indicate device or processing issues. There may also be outliers, such as zeros, in the data that typically indicate failures with the wafer sorting test itself.

Fig. 12.1 Wafer undergoing wafer probe. The three probes in the image have fine tips that contact to the bondpads of each device on the wafer. Once contacted, the wafer prober conducts electrical testing to distinguish good die from bad die. (Reprinted with permission from SST Components dba VPT Components [1])

From the wafer sort data, a wafer map of test values can be generated and used to examine wafer location dependencies. A radial distribution in the wafer map is common due to radial process non-uniformities such as in plasma processes. Non-radial patterns, such as spotty distributions, can indicate problems in the fabrication process.

In wafer probing, the wafer prober lifts the probes, steps the wafer to the next die, sets the probes down, and then tests the die. This process is prone to several failure modes. One example is that the wafer prober does not have infinite precision in its step, and as it moves across the wafer, its position can drift so that the probe tips no longer contact the die's bondpads. To accommodate this drift, the bondpads should have a minimum size. For MEMS devices, typical minimum pad sizes are on the order of 100 um × 100 um. Smaller bondpads are possible, depending on the capabilities of the prober, and would require identifying the wafer prober that will be used and confirming its capabilities during the die design stage.

Another example is when a group of probes contact the bondpads over and over again, they can scrape some of the conductive pad material, which then accumulates on the probe tips, causing them to short to each other. In the test program, it is useful to have a check for debris on the probes. A simple check is to raise the probes and check for electrical opens.

The duration of a wafer probe test will factor into the recurring cost of test. If there are, for example, 10,000 die on a wafer, and each die requires 10 s to test, then it will take approximately 28 h to test the entire wafer. This would be an unreasonably long time from a manufacturing throughput point of view. If one multiplies this duration by the 25 wafers in a lot, then it would take a month to completely wafer probe the lot, assuming only one probe station were available. A wafer sort time of a few hours per wafer would be considered more reasonable, which requires that each die on the wafer be tested within 1–2 seconds.

To improve test throughput, it is possible to test many devices at once. For example, a wafer probe head or probe card that can contact and test a 10 × 10 array of die can reduce wafer probing time by a factor of 100. To take advantage of this, the die and wafer layout should be arranged to reflect this pattern; for example, the die layout should be organized in 10 × 10 groups. It is also important to realize that the die may be electrically connected through the silicon wafer substrate. To enable parallel wafer probing, the die must either be designed to have electrical isolation from the substrate, or the wafer probe test must tolerate the substrate acting as a common node between devices.

Finally, the wafer sort data that is stored for each die should be accompanied by metadata. This includes, for example, die ID, wafer prober ID, operator ID, test date, test time, test condition flags, gage repeatability and reproducibility (gage R&R) data, temperature data, contact test results, plus an open field to allow the operator to record notes, such as unusual conditions. The metadata will be useful for data analysis and interpreting the wafer sort data.

Wafer Thinning

Wafer thinning may be needed for some MEMS devices. Typical 150-mm and 200-mm silicon wafers range in thickness from 600 to 800 um. However, usually only the top tens of microns of a wafer are used for the MEMS device. The remaining thickness of the silicon wafer is not necessary for the function of the device. To slim the thickness of the final product, the wafer can be thinned to as little as a few tens of micrometers. This can add value to a product where space is at a premium, such as in smartphones.

Wafer thinning is a chemically and mechanically intensive process. The substrate material to be removed is ground down with a slurry. The grind is often followed by a polish step. The MEMS wafer is mounted facedown to a tape, which can protect the MEMS devices from the water and the slurry, and the wafer is thinned from the backside. This process requires that the MEMS devices tolerate both the tape and grind process. Adhesive from the tape can damage delicate MEMS structures. One option is to recess structures so that the tape does not adhere to them. Another is to encapsulate the MEMS devices with a material that can be removed after wafer thinning, such as photoresist. Yet another solution is to seal a MEMS wafer at the wafer level by bonding on a cap wafer.

In addition, the MEMS die and wafer layout must be designed to withstand the grinding and/or polishing. For example, there should not be features that protrude from the surface of the wafer. The wafer itself must have enough structural integrity to not fracture under the grinding head. Wafers having large area cavities, for example, can be susceptible to fracture under these forces.

Singulation

After wafer sort and thinning, the wafer is singulated into individual die. The four most common methods are:

- Saw dicing
- Stealth dicing
- Plasma dicing
- Scribe and break

Saw dicing is the most common and most economical method. In saw dicing, a circular saw saws through the wafer (See Fig. 12.2). Saw dicing uses a jet of water to cool the saw blade, so the MEMS die must be able to withstand the presence of water and the pressure and flow of the jet. To protect the MEMS structures, the die can be either permanently encapsulated, such as with wafer-level packaging, or temporarily encapsulated, such as with adhesive dicing tape.

On the photomask, space between die must be added to accommodate the saw's path through the wafer. This space between die is often referred to as the "street" or

Fig. 12.2 Saw dicing of a wafer. The wafer is held by a tape, typically facedown to protect the MEMS devices on the wafer surface. A circular saw cuts through the wafer, and a jet of water is used to cool the cutting process. MEMS die that undergo saw dicing must be designed to tolerate the dicing process, including the tape application and removal, mechanical vibration, and the water jet. Reprinted with permission from Disco Hi-Tec Europe GmbH [2]

"lane." The street is typically 50–100 um wide. For small MEMS die, such as microphones, the wafer streets could take up a significant fraction of the wafer area. For example, for die that are 1000 × 1000um, the street could add 100um in each linear dimension. In this case, the streets would occupy ~17% of the wafer area.

With saw dicing, test structures can be placed within the street to save wafer area. The saw dicing would then cut through these test structures. In other dicing methods, test structures might interfere with the singulation process.

Another singulation method is stealth dicing. This method uses a laser to structurally weaken the wafer by creating a modification layer along the die perimeter. The die are separated by applying tension to the dicing tape, stressing the wafer along the laser weakened lines, and fracturing the wafer along the die perimeters. The laser used in stealth dicing does not ablate the silicon, and thus avoids the slag and rough edges of laser cutting. Stealth dicing is mechanically gentle and completely avoids the use of water or slurry, and thus is an excellent choice for delicate MEMS structures. Also, the kerf of stealth dicing is virtually nil, so very little wafer area is lost to dicing streets.

However, stealth dicing costs about 10 times as much as saw dicing, and therefore may be cost prohibitive for some products. Furthermore, stealth dicing is difficult to do on very small die (<1 × 1 mm). The adhesive force holding the die to the tape scales with die area and can be a limiting factor; it must be greater than the fracture force for singulation to occur during tape stretch. In addition, the laser must have a clear optical path to the silicon substrate in order to be able to structurally weaken it. Metal cannot be in the laser's path. Using stealth dicing therefore requires understanding of the stealth dicing process and planning during MEMS process integration and mask layout.

Plasma dicing uses a plasma to etch the silicon dicing lanes in order to singulate the die. Like stealth dicing, plasma dicing does not require water or slurry, so it is a mechanically gentle process. However, the plasma is chemically reactive, so the MEMS device must either tolerate the plasma or be temporarily covered, such as with photoresist. The width of the dicing streets will depend on the wafer thickness to be etched and the achievable etch aspect ratio of the plasma dicing tool.

The last singulation method is scribe and cleave. This method takes advantage of the single-crystal nature of silicon wafers. A diamond scribe dragged along the wafer surface introduces a flaw in the crystal, and then an applied force breaks the wafer along a crystal plane aligned with that flaw. This is a crude and unreliable technique because small variations may cause the cleave to deviate from the scribe line. Moreover, this is not a scalable technique, and is typically used only in the development phases to singulate a few die for initial evaluation.

If the scribe and cleave method is to be used, the die layout must be aligned to the crystal plane of the silicon. The flat on a (100) wafer marks one crystal plane, and the wafer will cleave both parallel and perpendicular to the flat. Note that the flat is cut typically to within ±2 degrees of the actual crystal plane. Intentionally introduced structural weaknesses in the wafer, such as partially etched streets, will improve the results of this singulation technique.

Package and Assembly

After the die are singulated, they are individually assembled and packaged. The package is the interface between the MEMS die and the system and the environment. (Chaps. 4, 9, and 13 discuss packages in further detail). There are many types of packages (see Table 4.3). To give a sense of the variety, packages differ by their materials, which are commonly ceramic, metal, and plastic. Packages can enclose single die or multiple die. Packages have a variety of shapes such as overmolded packages, cavity packages, and many others. With so many different packages, there are also many packaging assembly flows. Descriptions of all available MEMS packages and their assembly flows are beyond the scope of this book; instead we refer the reader to references for package and assembly [3, 4].

Designing a MEMS chip requires consideration of both the package type and its assembly flow. To illustrate the impact of a package and assembly flow on MEMS design, we choose an example, a cavity package and its assembly flow, shown in Fig. 12.3.

The process starts with a cavity package, which is a package having a space in the middle (the "cavity") where the MEMS die will be placed.

The term "die attach" refers to both the adhesive material and the process that attaches the die to the package. The MEMS die is attached to the bottom of the cavity by an adhesive which may be a liquid, commonly dispensed using a syringe, or a pre-cut tape. The adhesive serves several critical functions. It mechanically joins the MEMS to the package. It can also mechanically isolate the MEMS from any package induced strain by choosing a low modulus polymer material. As the package bends or expands, the soft polymer adhesive deforms, allowing the MEMS chip to be isolated from the package's strain. The die attach material can also be thermally and/or electrically conductive if the MEMS device requires it.

A robot is often used in the die attach process, which places requirements on the MEMS design. In the die attach process, a robot reads the wafer map to select a

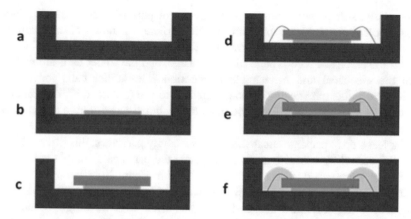

Fig. 12.3 Example assembly flow for cavity package. (**a**) Start: Cavity package. (**b**) Die attach adhesive dispense. (**c**) Die attach process and adhesive cure. (**d**) Electrical attach (wirebond) (**e**) Gel protection. (**f**) Package encapsulation

good die. The robot then applies a pin from the backside to push the good die off the singulation tape. If there are features on the back of the MEMS die, their design must accommodate this pin. A robotic arm then picks up the die, looks for alignment marks in the package, aligns the die, and places it onto the uncured die attach adhesive in the cavity of the package. Often the robot picks up the MEMS die by using suction. As a result, the die should have a flat area, ideally in the center, to allow the head to make contact and pull vacuum. For MEMS having fragile structures that could not survive such vacuum, such as suspended membranes on pressure sensors, the robot can handle the die by its edges.

Once the MEMS has been attached to the package, then the electrical connections can be made. The most common method is wirebonding, where a thin (~25 um diameter) wire is thermally and/or ultrasonically bonded between the MEMS electrical pads and the package electrical feedthroughs. Au and Al are the most common wire types used in wirebonding.

Using wirebonding requires bondpads to be of a compatible material (also usually Au or Al, respectively) and a minimum size and pitch to accommodate the wirebond tool head. Typical bondpad sizes for MEMS devices are 100 um × 100 um on a 200 um pitch, but both smaller and larger bondpads and pitches are possible. Bondpads typically need to be placed on the perimeter of the die. Finally, the wirebond length should be minimized and wirebonds should not cross over each other. This requires that the bond pads of the die and the bondpads of the package are near each other and match in sequence. Once the wirebonds are formed, they can be covered with a gel that protects the wirebonds from the environment or damage during handling.

An alternative method to electrically and mechanically attach the MEMS is to use solder bumps. The solder bump method is useful for situations needing many electrical connections, particularly if in a small area. While wirebonds are made sequentially, solder ball connections are made in parallel. There are a variety of ways to place the solder bumps on the die surface, including electroplating the wafer and reflowing the solder material to form balls. The MEMS and solder balls contact the electrical feedthrough of the package, and the solder balls are melted to form the electrical and mechanical connection. The solder balls form a rigid mechanical joint, transferring package stresses to the MEMS die. If a solder ball approach is selected, the MEMS die must be designed to be able to accommodate this stress environment.

The last step is package encapsulation. For cavity packages, this is done by adding a lid, sealed by adhesive or welding. Once the package assembly is complete, the packages are tested (discussed in more detail in Chap. 20).

The assembly flow discussed in this example is only one possible packaging approach, to illustrate how backend processes affect MEMS design. Some of the details discussed here, such as wirebonding, also apply to other types of packages. Other details are specific to cavity packages, such as lidding. Regardless of which packaging approach is used, it should be identified early in product development, and its requirements understood, so that the MEMS design is compatible with both the package and its assembly flow.

Summary

The back end is an essential part of the overall MEMS manufacturing process, and it imposes requirements on the MEMS design. The back end is composed of multiple steps and each step has multiple options. In turn, each option imposes its own set of requirements on the MEMS design. The final set of options will need to be decided with the MEMS device and package in mind and vice versa. The back-end process design should therefore be evolved with the MEMS design early in the development program so that each can be adjusted to accommodate the needs of the other.

References

1. Probe. SST Components dba VPT Components. http://www.solidstatetesting.com/test/ and http://www.vptcomponents.com. Accessed 29 May 2020
2. Dicing (Kiru)—Blade Dicing. DISCO HI-TEC EUROPE GmbH. https://www.dicing-grinding.com/services/dicing/. Accessed 29 May 2020
3. Lee YC, Cheng YT, Ramadoss R (2018) MEMS packaging. World Scientific Publishing, Singapore
4. Lau JH, Lee CK, Premachandran CS, Aibin Y (2010) Advanced MEMS packaging. McGraw-Hill, New York

Chapter 13
Strategies for Codevelopment of the Electronics and Package

A common mistake in the development of a MEMS or microsystem product is to exclusively focus on the development of the MEMS chip. While the MEMS is usually the distinguishing part of a MEMS product, the other main components, specifically the electronics and packaging, are necessary for the MEMS product to function, and moreover, they offer significant opportunities to increase a product's value.

The development timelines of the package and the electronics may be long, and one should not wait until the MEMS has been completed to then start the development of the other components. A typical lead time for the first iteration of a custom ASIC is 12–18 months. The lead time for a custom package is several months, potentially longer, depending on the requirements of the package (see Fig. 13.1). Given these long timelines, the development of the electronics and package should be done in parallel with the development of the MEMS.

Parallel development of the MEMS, package, and electronics significantly shortens the overall development timeline, but makes the development a more complex undertaking. The risks and costs of parallel development can be managed and mitigated with awareness of the issues and advanced project planning.

There are many good technical references on the separate development of each component. This chapter focuses on the particular challenges of codevelopment and offers useful strategies.

Codevelopment Challenges

The primary challenge of parallel development is that it requires the simultaneous development of three immature components which are strongly coupled. These components are so strongly coupled that none of them can be fully tested and

© The Author(s), under exclusive license to Springer Nature Switzerland AG 2021
A. M. Fitzgerald et al., *MEMS Product Development*, Microsystems
and Nanosystems, https://doi.org/10.1007/978-3-030-61709-7_13

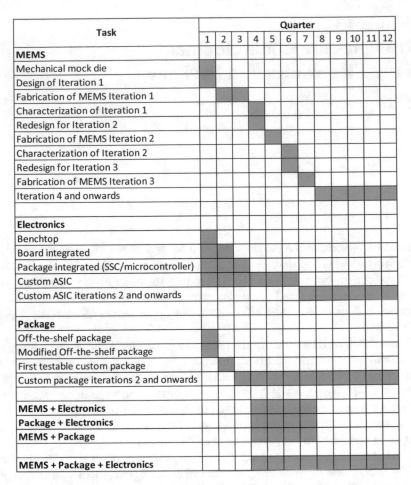

Fig. 13.1 Example Gantt chart for a MEMS product development program having parallel development of the MEMS, electronics and package

characterized without the other two. If all three components are under development, then how can any of them be characterized?

The general approach we recommend is to develop each component using proxies for the other two components. These proxies are created by first developing cheap, fast, and therefore easy-to-iterate embodiments before moving to the final embodiment. The speed enables testable embodiments to be brought quickly into the development program. Because they are fast to produce, the proxies are also fast to iterate, and so designs may be updated and bugs fixed quickly. When bugs are fixed quickly, confidence develops in these embodiments, and in turn, they become good proxies for the other components.

In addition, these proxies expand the capabilities of the development team. For example, once the electronics team develops a board for reading the electronics, multiple copies of it can be made, so that the MEMS and packaging teams could each have multiple sets of the electronics available for their own experiments.

Using Proxies to Enable Parallel Development

The MEMS device is usually the slowest component to develop, but there are several proxies for the MEMS that can be used to enable packaging and electronics codevelopment.

The earliest proxy for the MEMS is a virtual one. The MEMS may be represented by analytical equations or by models extracted from finite element simulations (for more discussion on simulations see Chap. 10). These virtual proxies may be used for simulations of the electronics and for development of software algorithms. The models of the MEMS device allow the electronics and software teams to begin their development before fabrication of the MEMS device begins.

The first physical proxy is a mechanical dummy die, which represents the MEMS and/or ASIC's form factors. Mechanical dummy die are simply blocks of silicon whose dimensions match the product die they represent. In addition, they should have bonding pads of the same size and location as the die they represent. Mechanical dummy die are useful for developing the back-end-of-line (BEOL) processes, such as the wafer probing, singulation, thinning, pick and place, and package assembly (discussed in more detail in Chap. 12). For example, dummy die can be used to develop the programming for the machinery that performs die attach and wirebonding, so that when the working MEMS die are available, these processes are already developed.

Another useful proxy is an electrical dummy. This is similar to the mechanical dummy, but has static electrical impedances that are in the same range as the functional MEMS die. A packaged electrical dummy can capture many effects that the package may have on MEMS electrical performance, for example, humidity testing of the package. The electrical dummy can help address questions such as, does the package sufficiently protect the MEMS chip from moisture? Are there electrical shorts? Is there galvanic corrosion? An electrical dummy can be packaged and then exposed to moisture to help answer these questions while the actual MEMS is undergoing development.

A MEMS proxy that is useful to a custom ASIC is an "internal electrical dummy," which are internal ASIC electrical loads that can be switched in and out of the ASIC's port to the MEMS. The ASIC team can advance its development with this internal proxy without the MEMS. Furthermore, the internal electrical dummy will later enable the ASIC to conduct self-testing once the product is in the field.

A MEMS proxy die may also have test structures which can provide information on mechanical stress, electrical impedance, chemical resistance, and more. One example of a simple test structure is a dummy die with the pads shorted. This is useful for developing wirebonds that must meet product qualification criteria. For example, a common qualification test requires that the system survive 5 drops onto a cement floor from a height of 2 meters in the $\pm X$, $\pm Y$, and $\pm Z$ orientations. A common failure of this test is the mechanical shock of the drop detaches the wirebonds from the bondpads and creates an electrical open. This detachment can be difficult to detect visually, as even a microscopic separation will create an open. Rather than attempt visual inspection, this failure mode is more easily detected electrically. The shorted pad dummy die is wirebonded in a package which is then subjected to the drops. Afterwards, broken electrical connections can be found by checking electrical continuity.

Another test structure may be used for detecting spurious mechanical stress from packaging. Plastic packages are attractive because they have low per-unit cost, but they are challenging to implement for some MEMS devices. Of the three common packaging materials, ceramic, metal, and plastic, plastic has the largest coefficient of thermal expansion (CTE) mismatch to silicon. Plastic also experiences creep, hysteresis, stress relaxation, and other phenomena that present engineering challenges to the MEMS device. These effects can be characterized with test structures in packaged die while the MEMS is still under development [1].

An example stress test structure is an angularly amplified pointer [2] whose end moves as the anchors experience displacement due to stress on the die from the package. This die can be assembled in the package and the stress on the die observed with a microscope, using a hotplate to provide thermal stress.

Electronics Development Proxies

The first physical embodiment of the electronics is typically an implementation via benchtop instruments, breadboard circuits, and software, such as National Instruments' LabView. This embodiment is quick and simple to set up and alter, but each instrument may be expensive. Sometimes, there may be only a few of these instruments available, so test setups must be repeatedly assembled and broken down by the engineers in order to share the instruments. This slows testing and introduces errors in the setup.

Other proxies for the electronics follow the levels of integration described in Chap. 9. At the PCB level, the electronics are composed of off-the-shelf components. The build for a PCB costs in the hundreds of dollars and typically takes a few days to a few weeks to receive. With costs and lead times of this order-of-magnitude, updates to the design are quickly and easily made. Once the design is finalized, many copies of the board can be made, so that electronics capability for the MEMS is abundantly available to the development team.

At package-level integration, the electronics and the MEMS share the same package. Package integrated embodiments require that the electronics be integrated into a single or a few electronic components. This can be embodied by a sensor signal conditioning (SSC) chip, a microcontroller, and/or a custom ASIC. A discussion of SSC and microcontrollers is below. With a package integrated system, every MEMS device and every package has its own set of electronics.

Die-level integrated electronics are the most integrated form of the MEMS product, and so the other levels of integration may serve as proxies for die-integrated microsystem development.

Whatever the final level of product integration, useful proxies can be made along the way. For example, if the final product will be a package-level integrated microsystem incorporating a custom ASIC, then a PCB embodiment of the ASIC design serves as an iteration for debugging the system electronic design, and a final version of the PCB serves as a proxy for the ASIC in the MEMS and package development.

Leveraging Sensor Signal Conditioning (SSC) Chips and Microcontrollers

Two powerful components that can serve as electronics proxies are the sensor signal conditioning (SSC) chip and the microcontroller. Both are available off-the-shelf and can provide many of the functions needed in MEMS applications quickly for either the PCB or package-level integrated system.

The SSC serves as the electronic interface from the MEMS chip to the rest of the product system, and includes in its functionality the full signal conditioning path, including signal amplification, filtration, analog-to-digital conversion, calibration, and compensation [3, 4]. The SSC may be capable enough that it could serve as the electronics embodiment for the final product.

Another useful component during electronics development is the microcontroller. The microcontroller is a more general-purpose chip than the SSC, but does not usually have the SSC's signal conditioning capabilities. Along with basic interface capabilities to sensors and actuators, the microcontroller also has a programmable CPU, short-term memory (similar to a computer's RAM), and long-term storage (similar to a hard drive). The presence of a microcontroller is akin to having a computer on the board or in the package, and it is used to run many software-based functions, such as high-level feature extraction, system self-monitoring, and data logging during testing.

Additional capabilities that can be found in microcontrollers include analog-to-digital converters (ADC), digital-to-analog converters (DAC), on and off chip temperature sensing, internal timers, standard digital interfaces (e.g., SPI, I²C), wireless capability, and many more. See Chap. 9 for further discussion of microcontrollers and software in a MEMS product.

The programmability of the microcontroller allows it to be used as a testbed for algorithm development for the MEMS device. Estimating algorithm resource requirements, such as computation power, RAM, and storage space, can be difficult early in development. In addition, over the course of development, new functions may be identified, and the number of algorithms often increases. Developing the algorithms on a microcontroller allows the algorithms to approach final form and the algorithm resources can then be more accurately estimated for specifying a custom ASIC design.

The microcontroller's presence during development also broadens test capability. For example, its flexibility enables it to act as a datalogger during testing. Rather than having a single test system testing 100 devices sequentially, a group of microcontrollers coupled directly to the MEMS can take data in parallel. Moreover, the microcontroller's size and power consumption make it useful for wireless and battery-operated testing, and can be used to monitor the system during product qualification tests such as drop and vibration testing.

Since a microcontroller's signal conditioning capability is not usually as sophisticated as that found in a SSC, a powerful combination is to use both chips together to realize a system that has the programmability of a microcontroller and the signal conditioning sophistication of a SSC.

There are many microcontrollers commercially available. They vary in computational capability, wireless communication capability, power consumption, instruction set architecture, development tools, software libraries, and many other factors. The microcontroller used during development should have the same instruction set architecture as that to be used in the final product. This will smooth porting of algorithms written during development to the final product. Often this means choosing an open source or licensable microcontroller architecture that can be integrated into a custom ASIC.

A consideration for package level-integrated systems is identifying SSCs and microcontrollers that are available as an unpackaged part (also known as "bare die"). Most ICs are only available as packaged devices. But there are manufacturers that sell their SSCs and microcontrollers in bare die form, which allows their inclusion into package-integrated products.

System-Side Electronics Interfaces

Electronics interface between the MEMS and the greater product system. There are at least two systems that the electronics must be able to interface to. One is the customer's, and the other is the MEMS company's test stations used during development. By defining a common interface to both of these systems throughout the development program, proxies for the electronics may be developed without duplication of engineering effort.

To illustrate, suppose the final product is package-integrated with a custom ASIC. One interface is between the electronics and the MEMS. The other interface is defined between the electronics and the rest of the system. See Fig. 13.2. The product is composed of the MEMS and the electronics, and they interface to the customer's system. During the development program, the MEMS must also interface with characterization and testing systems via the electronics, which may be embodied as either a PCB or an ASIC. If interfaces are maintained consistently throughout the product development, then different embodiments of the electronics can be substituted whenever they are ready without substantial re-engineering at the interface. And if they can be substituted, then they may act as proxies for each other during development of the MEMS product.

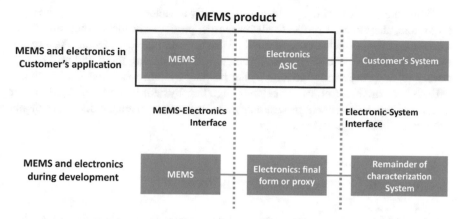

Fig. 13.2 Interfaces for the electronics should be consistent for both the customer's system as well as for the characterization systems used in development

Packaging Development Proxies

Off-the-shelf packages may be used as a proxies for codevelopment, and there are many different types to choose from [5, 6]. One basic choice is the package material, which includes ceramic, metal, and plastic. Each material has its advantages and disadvantages, and they each have different purposes in the development stage. For example, a metal package could isolate the MEMS die from electrical noise. A ceramic package is the closest CTE match to silicon and can be used to isolate thermal effects. A plastic package can simulate some mechanical effects of a custom plastic package, such as mechanical shock transmitted to the MEMS through the package.

Off-the-shelf packages usually need to be modified for use with MEMS chips. Most packages are intended for CMOS chips; almost all MEMS devices require a port to interact with the phenomena they are sensing/actuating, such as sound, pressure, and chemistry. The cost and time to modify a package has a wide range. It can be as simple a hole drilled into the package to allow a physical stimulus like pressure to interact with the MEMS. It can also be a complicated modification, such as a package for resonant micromirrors. This package would require a transparent, optical window that is also hermetic to enable operation in partial pressure. This level of alteration could cost thousands of dollars and take weeks to implement.

Custom packages are similar in development complexity to custom MEMS or electronics, in that they have an in-depth NRE phase for design and prototype fabrication, tooling setup, and the fabrication run. Custom packages typically take months, up to a year, from initial request to receipt. Moreover, the custom package will likely need to be iterated several times in order to finalize its design and manufacturing process. The order of magnitude NRE cost for a custom package is tens of thousands of dollars to hundreds of thousands of dollars.

While the proxies for the package can usefully advance development of the overall system, they cannot replicate all responses, particularly the second- and third-order interaction effects between the final package and the MEMS die. As a result, even with strategic use of proxies throughout development, there will still be discovery of unforeseen interactions in the product system when the final package, MEMS, and the electronics are brought together. System integration should occur as early in the development program as possible in order to enable discovery of any unforeseen interaction effects.

Examples of Codevelopment with Proxies

As the MEMS, package, electronics, and their respective proxies proceed through iterative development, they may be utilized in different combinations by each development team at various points in the project. See Table 13.1. Not every product development program will need to develop all of these proxies, and there is no fixed recipe for which combinations to use when. Usage will depend on factors such as the availability and maturity of the proxy and the particular goal of the team at a particular stage of the development. We give some examples below to illustrate how proxies may be used in parallel development.

When a MEMS development team is characterizing an early prototype MEMS die, they will likely have the MEMS in a modified off-the-shelf package, read using benchtop instruments. They would choose this combination because their focus is on evaluating the MEMS die, which, being the first iteration, will likely have many issues and need de-bugging. As a result, the MEMS team would like to experience as few confounding effects from the package and electronics as possible. So, they would choose to use the most mature, most risk-free options: a modified off-the-shelf package and benchtop instruments. This allows them to focus analysis efforts solely on the MEMS and not have to debug the package or electronics.

When the electronics or test team is developing the first board, they may work with an electrical dummy of the MEMS in an off-the-shelf package, or they may work with no dummy and simply use static impedances, such as an off-the-shelf

Table 13.1 List of proxies used in different stages of MEMS product development

Stages of development	MEMS	Electronics	Package
Proof of concept	Mechanical dummy	Benchtop instruments	Off-the-shelf package
Advanced prototypes	Electrical dummy	Board-level integrated	Modified off-the-shelf package
	Dummy with test structures	Package-level integrated	
	MEMS-iteration 1	Custom ASIC-iteration 1	Custom package-iteration 1
	MEMS-iteration 2	Custom ASIC-iteration 2	Custom package-iteration 2
Production	MEMS-iteration 3	Custom ASIC-iteration 3	Custom package-iteration 3
	⋮	⋮	⋮

resistor. Like the MEMS team, the electronics team would choose this combination because these proxies are the least likely to introduce spurious issues, and therefore the electronics team could focus on isolating and resolving any issues with the new electronics design.

The package team may elect to work with a mechanical dummy for both the MEMS and the ASIC, in order to develop the back-end-of-line process. This proxy allows the packaging team to program the wafer prober, test the probe card, program the pick and place machinery, set wirebonding parameters, and so forth.

The software team may start with virtual models of the MEMS device implemented in code. Models allow the software team to start algorithm development before the electronics or MEMS teams have produced any hardware.

Later in the product development, the MEMS and packaging team work together with a latest iteration of the MEMS die and the latest iteration of the package. In order to investigate MEMS-package interaction effects, they may use board-integrated versions of the electronics, because they are now sufficiently debugged and reliable. Since many copies of the boards can be cost effectively produced, the MEMS team could test many aspects of the MEMS chip-to-package interactive effects, such as temperature response and humidity response. These tests could be executed in parallel at several different test stations.

The test team, meanwhile, would like to scale up the test capacity. They would use an electrical dummy as a simple MEMS proxy, and a package having the same pinout as the custom package. To decrease the individual device test time, they are developing a test station that could test 30 devices simultaneously. Because the ASIC is not yet ready, the new test station uses 30 electronic proxy boards to read the devices under test.

The electronics team is evaluating the first iteration of the custom ASIC. For the MEMS proxy, they use the internal electrical loads inside the ASIC. These impedances are well known and can be conveniently switched to evaluate the input impedance range and sensitivity of the ASIC.

The software team could use a board with the latest MEMS chip mounted in a modified off-the-shelf package in order to develop algorithms that allow higher-level feature extraction from the MEMS sensor data.

Towards the end of the product development, a final iteration of the microsystem that incorporates the final versions of the MEMS, package, and ASIC would undergo product qualification testing. One hundred of these units would need to be exposed to heat and humidity cycling while powered. Wired power and communications is difficult and expensive to implement with so many devices subjected to such adverse environmental conditions. So each ASIC's microcontroller functionality is instead programmed to act as a datalogger and record the output of each sensor during the test.

Finally, the test team develops a manufacturing test system that needs to test, calibrate, and compensate 10,000 devices per week. The test system conducts a trial run with 500 known good units of the product, which consists of the final package, ASIC, and MEMS. The test team uses these units to check the new test system's device handling, electrical interface, and test throughput.

Summary

MEMS products have three key components: MEMS device, package, and electronics. Parallel codevelopment can shorten the overall development timeline, but has greater complexity due to the fact that the MEMS, package, and electronics are undergoing development simultaneously. Proxies for each of these parts enable and facilitate parallel development. Strategic use of proxies in the overall development program can accelerate product development, reduce cost, and enhance development capabilities.

References

1. Cassard JM, Geist J, Vorburger TV, Read DT, Gaitan M, Seiler DG. *Standard Reference Materials®* User's Guide for RM 8096 and 8097: The MEMS 5-in-1, 2013 Edition. National Institute of Standards and Technology Special Publication 260-177
2. Masters N, de Boer M, Jensen B et al (2001) Side-by-side comparison of passive MEMS strain test structures under residual compression. Mechanical Properties of Structural Films. ASTM International
3. Fraden J (1993) Interface electronic circuits. In: AIP handbook of modern sensors: physics, designs and applications. AIP Press, New York
4. Bryzek J, Petersen K, Mallon JR, Christel L, Pourahmadi F (1990) Signal conditioning for sensors. In: Silicon sensors and microstructures. Lucas NovaSensor, Fremont
5. Lee YC, Cheng YT, Ramadoss R (2018) MEMS Packaging. World Scientific Publishing, Singapore
6. Lau JH, Lee CK, Premachandran CS, Aibin Y (2010) Advanced MEMS packaging. McGraw-Hill, New York

Chapter 14
Planning a Development Test Program

MEMS design and fabrication usually receive the focus in a development program, and too often, testing is treated as an afterthought. However, the test phase of the design-fab-test iteration can be the most time-consuming part because it requires debugging and failure analysis, whose durations are unpredictable. With engineers' salaries comprising a large part of a development program's budget, the most time-consuming task is often the most expensive task.

In addition, the test equipment is usually specific to a MEMS development program, and if it cannot be purchased, then it must be developed. This is a substantial investment of time, money, and engineering resources. Moreover, the development of test equipment must also go through its own design-build-test iterations. With some planning and consideration, a well-thought out test program can reduce the overall cost, time, and risk of MEMS product development.

"Test" is a broad term and it can refer to many different types of testing (see Fig. 14.1). For the purposes of this book and this chapter, we use the term development test to refer to testing that occurs during the development phase, whose goal is to inform the product design and process on how the device succeeds or fails to meet its performance specifications, and how the design and process need to be improved. Development test includes characterization or performance test of the device, such as measuring its sensitivity or range. Development test also includes product qualification testing, which tests the product's performance and reliability under operating conditions. Failure analysis is also part of development testing, and its goal is to understand why a device design and process might be failing to meet requirements. Finally, manufacturing test is used to refer to testing during the manufacturing phase in order to filter bad parts and to control product quality.

A. M. Fitzgerald et al., *MEMS Product Development*, Microsystems and Nanosystems, https://doi.org/10.1007/978-3-030-61709-7_14

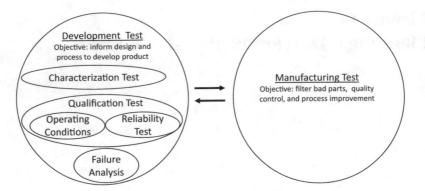

Fig. 14.1 The relationships between various types of testing used in MEMS development and manufacture

Table 14.1 Differences between development test and manufacturing test

	Development test	Manufacturing test
Test volume	Few devices	Many or all devices
Test data	Thorough characterization	Minimal characterization
Test time	Minutes to days	Seconds

Understanding Test Types: Development Test vs. Manufacturing Test

First, let us clarify that test falls into two broad categories: test for product development and test for product manufacturing. Each has different objectives. Development test's objectives are device characterization, product qualification, and failure analysis. Manufacturing test's primary objective is identification of nonfunctioning or substandard parts and their removal from the manufacturing process. Secondary objectives include process improvement of the manufacturing process and binning, which is the practice of grading functional devices by their performance so that they may be sold as different products.

Some high-level differences between development test and manufacturing test are summarized in Table 14.1.

Development and manufacturing test are complementary and inform each other. For example, failure modes characterized during development test often identify which tests will be most useful in manufacture test. And vice-versa, manufacturing test can scan a large number of devices to identify a population of defective devices which are then more thoroughly investigated at development test.

This chapter focuses on development test. Manufacturing test is discussed in Chap. 20.

Development Test

During MEMS development, design, fabrication, and test form an iterative work-flow loop (Fig 14.2). A device is designed, fabricated, and then tested to see how it performs. Except for the final iteration, the testing will discover shortcomings or failures in the design and/or process. The failures are analyzed and understood. Then the design/process is updated to correct these failures. The new version is fab-ricated, tested, and iterated until a version that passes all qualification tests is real-ized, and the product achieves "product qualification." When product qualification is attained, the product can move from development to manufacturing.

It should be emphasized that product qualification applies to the entire MEMS product, and not just the MEMS die. If, for example, a MEMS product includes both the MEMS die and its package, then the packaged MEMS die must pass prod-uct qualification.

In an iteration loop, failure analysis can be the step with the biggest variance in duration. Some failures are quick and easy to understand. Some failures are mad-deningly difficult and time consuming to figure out or even to replicate. Regardless, a prototype failure needs to be understood in order to identify design and process changes to address the failure.

How Testing, Performance, Reliability, Product Qualification, and Failure Analysis Are Related

Within development test, we identify several types of testing: characterization test, reliability test, qualification test, and failure analysis. These are terms that will be used in this chapter and our definitions for their use in this discussion should be explained.

Fig. 14.2 Typical workflow for MEMS product development

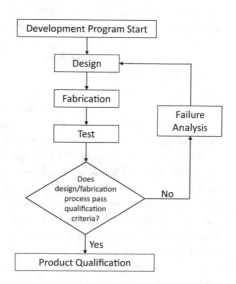

Characterization test is the testing of the product to its product specifications, also called performance specifications. An example product specification for an absolute pressure sensor may be a sensor for tire pressure monitoring system (TPMS) that has a range of 0-50 psi (a typical tire pressure range) with part per thousand accuracy.

The MEMS product will need to be tested under external conditions, known as its operating environment. Since the above pressure sensor is intended to be used for TPMS, then the operating environment would include conditions such as temperature, humidity, water, mechanical shock from road bumps, and exposure to gas oline.

Product qualification is a gatekeeping event which demonstrates that the device meets the qualification requirements set by product definition. For the TPMS pressure example above, a qualification requirement may include that the pressure sensor meets its product specification, i.e., its range and accuracy, over a temperature range from −40 to +120 °C.

Qualification test subjects the product to the qualification requirements in a structured, documented way. These requirements commonly include demonstration of the performance specifications under operating conditions, and include reliability testing. An example of a reliability test is the Highly Accelerated Stress Test (HAST), which is a standard test and is discussed in more detail below. The product is subjected to the HAST test and then characterization tested afterwards to demonstrate that the product still meets specification.

Despite product qualification being a one-time event, the set of qualification tests may be done many times during development, as new iterations of the device design and fabrication process are tested against the qualification criteria. The nature of development is such that the prototypes will fail the qualification tests again and again, and the device design and process would be iterated until a design and process are able to produce a version of the product that can meet the qualification criteria.

When products fail the qualification tests, failure analysis identifies how and why they fail. Failure analysis is needed to understand how to update the device design and process in the next design-fab iteration. With the pressure sensor example, suppose the pressure sensor is subjected to the HAST conditions, and afterwards, characterization test shows that the sensor fails the part per thousand accuracy requirement. The sensor then goes to failure analysis to understand how and why it failed. Perhaps thermal stresses in the package shifted the pressure response curve. Perhaps the wirebonds in the package developed increased contact resistance. These are different failure mechanisms, and therefore they require different changes to the device design and/or process to address the failure. Once the failure is understood and the design and process have been updated, the product is fabricated again, and the updated version is again tested to see if the problem has been fixed.

Once the product meets all of the qualification requirements, the design and process are frozen and the product moves to manufacturing. After qualification, further design or process changes may still be possible, but may also require detailed regulatory documentation or communication and possibly negotiation with customers. If extensive changes are needed, then the product may need to be re-qualified. Investing time to discover all product and customer requirements at the start of product development (see Chap. 4) will help avoid costly repetition of qualification.

Establishing Test Requirements

The first step to planning a test program is to identify the testing requirements. These derive from the product's performance and qualification requirements. It is important to neither underspecify nor to overspecify the test requirements. Underspecifying the requirements will result in a device that could later fail during its operation. Overspecifying requirements will incur unnecessary engineering time and cost.

The product's performance and qualification requirements derive from the product definition. To identify the test requirements more fully, consider what may happen to the device from the point it leaves the company's production facilities to the end of the device's life. This means taking into account conditions beyond normal operation, including shipping, storage, assembly, repair, as well as potential product abuse and misuse.

Standard Tests can Leverage Existing Equipment and Expertise

There are number of tests that have been defined by professional standards organizations. Standardized tests are well defined and well understood. There are experienced people who can be consulted on the tests' methods, design, and specifics to help the development team identify which tests to apply. Test machinery that conducts standardized tests is often available for purchase, rental, or as a service.

Standardized tests can also be helpful in translating requirements into specific testing procedures. For example, a customer requires that the product is reliable and meets its specifications throughout its lifetime. How does this translate to a set of tests, particularly if the product requires a long lifetime? A good place to start is existing reliability testing standards (see Table 14.2). These include tests for vibration, humidity, temperature, temperature cycling, and many more. One example is the Highly Accelerated Stress Test (HAST) [1]. There are different versions of the HAST test, but the version from JEDEC, a standards group for the microelectronics industry, is commonly used in MEMS. Even within the JEDEC HAST test, there are multiple versions. The version that is most applicable will depend on the product as well as its application. One version of the JEDEC HAST test requires subjecting a sample of products to the following conditions:

- Temperature at $130 \pm 2°C$
- Relative humidity at $85 \pm 5\%$
- The devices are at maximum power
- The devices are subjected to these conditions for 96 h

Some of the methods that HAST employs include applying high temperature to accelerate processes that may degrade the device, such as material aging. The humidity is used to simulate processes such as water penetration of cracks. Powered operation simulates electrical effects such as galvanic corrosion.

Table 14.2 List of standards organizations and the type of standards they issue relevant to MEMS

Standards organization	Type of standard
AEC	Automotive
ANSI	General
ASTM	Mechanical engineering
FDA	Medical
ISO	General
JEDEC	Microelectronics
MIL-STD	Military
SEMI	Microelectronics

It is important to understand the purpose of the test and its methods. The standard test should not be blindly applied to your product. For example, if the standard test requires subjecting the product to 130 °C, but the product melts at 110 °C, then that test does not make sense for your product. Consult with experienced specialists who can advise on which tests are appropriate and and how they can be used to improve your product.

The Role of In-Field Testing in Defining Qualification Tests

Characterization and qualification tests are only proxies for conditions that the device will experience in the real world. It is important to keep in mind that there is no perfect substitute for actual real-world use. As a result, in-field testing must be part of any development program. However, in-field testing can be expensive, slow, uncontrolled, and too resource intensive to be the main testing method during device development. For example, testing a new tire pressure monitoring system (TPMS) in an actual tire on a car is valuable, but it is difficult to reproducibly subject the TPMS to the required range of temperatures, humidities, shock, etc. Instead, for certain end use applications, in-field testing may be used as a way to identify characterization and qualification tests.

An example is the development of TPMS. In-field testing on a real tire might reveal that the TPMS returns an error when the automobile accelerates. With this realization, tire rotational acceleration conditions may be added to the in-lab tests of the TPMS product.

Establishing Test Capability

"Test capability" refers to the hardware and software needed to conduct testing. It includes the test equipment, test electronics, test fixtures, software, data analysis, algorithms, etc. We use the term "test volume" to refer to the number of devices that the test can characterize at once, and "test speed" or "test time" refers to the throughput or duration, respectively, needed to complete a set of tests.

For many MEMS devices, test capability for a development program is often specific to the device and the application. The test capability needed for a pressure sensor in a tire pressure monitoring system is different than the test capability for an accelerometer for a smartphone. As a result, test capability cannot usually be purchased off-the-shelf. It must be developed in-house or subcontracted to a test development group. As a result, establishing test capability is a major investment for the MEMS product development program. Understanding how test capability is developed will enable thoughtful test program planning and strategic and efficient use of resources.

Characterization Test is a Core Capability

The characterization test hardware is the core of the test capability. This hardware will test the product's performance specifications over and over again. Because there are many different MEMS devices for many different applications, we will use one to illustrate the considerations that need to be taken into account in developing the characterization test equipment. For this purpose, we will continue to use the TPMS pressure sensor as an example.

Once the test requirements have been identified, the test equipment capabilities can then be defined. For a pressure sensor, this includes the ability to measure the pressure sensor's range, accuracy, as well as many other performance specifications, such as hysteresis, drift, and nonlinearity.

A characterization test station needs to apply stimulus inputs to sensors and read outputs. Moreover, the test station must apply the input stimulus over a wider range and higher accuracy than the product to be developed. To illustrate, say that the TPMS pressure sensor is required to have a range of at least 0-50 psi with an accuracy of part per thousand. A test station must apply pressures with a range wider than 0–50 psi and with an accuracy greater than part per thousand. Developing such a test station can be a substantial undertaking. Fortunately, highly accurate pressure controllers over this range are available off-the-shelf, and may be purchased for a characterization test station.

The characterization test station must also supply power to the MEMS and read the output of the sensors. The interface electronics are important part of not just the test station, but the entire development program. There are advantages to consistently defining the electronics interfaces so that the sensor will be able plug into both the test system and the customer's end application (discussed in more detail in Chap. 13 on MEMS package and electronics co-development).

For sensors, the characterization test station should have flexibility to apply different test profiles. Again using pressure sensors as an example, some tests may need to sample at several different pressures in order to create detailed performance curves. Other tests may need to quickly sample at the minimum and maximum operating pressures. Yet other tests may need to accurately hold one pressure continuously in order to measure device stability over time.

Because characterization testing will need to be performed repeatedly and frequently throughout the development program, the characterization test station should be capable of enough test volume and test speed to be able to run the most common test profiles within a few hours at most. In addition, characterization test stations should be continuously available. The development team should *not* "borrow" equipment from a test station to set up another test. A disassembled test station must be reassembled, which introduces errors in the test setup. Ideally, the characterization test station should be as automated as possible, so that the development team can focus on interpreting the results of the test and not running the test.

All test stations, not just the characterization test station, should be able to record metadata that will be essential for tracking and interpreting results. Metadata include the date and time for each reading, the test station ID, the test operator, software versions, and an open field that allows the operator to append notes on the test, such as the test objective and unusual observations. Environmental variables such as temperature and humidity can also be useful information, which requires that the test station have the capability to monitor and log those parameters.

A real example of the utility of metadata: a batch of pressure sensors under development had an intermittent signal where the sensor output temporarily rose and then settled again. The effect was small, but enough that the devices failed their accuracy specification. The duration of this elevated signal was tens of minutes to hours. Weeks of puzzling were spent trying to figure out the issue. Finally, after checking the time/date stamps of the metadata, it was realized that the intermittent signal coincided with the building's HVAC system cycling on and off. The test station was upgraded to isolate the devices from the building's HVAC influences which removed the intermittent error.

Finally, having a database to record and organize all test results allows the team to focus on test analysis instead of data management. There will be lots and lots of test results. If they are recorded in different files on different computers, it will be difficult to collect and analyze them. A database of test data that is searchable by the part identifiers, design version, fabrication process version, test conditions, metadata, etc. will greatly amplify the development team's ability to analyze product performance and conduct failure analysis.

Developing Characterization Test Capability

The characterization test station is a complex machine that, like the MEMS product, requires its own development iterations of design-build-test. Fortunately, the test capability can grow over time, and early iterations can become prototypes for later iterations, and even test stations themselves.

Early on, typically only a few MEMS devices yield from fabrication, so there are only a few MEMS devices to test. At this point, a test station that can characterize a single device is adequate. This test station should prioritize flexibility over test speed. The early MEMS prototypes will have a lot of problems, and being able interrogate the device in different ways will be useful in its failure analysis. Again, using the TPMS pressure sensor as an example, in addition to applying different pressures, it may be useful to be able to test the device with different voltages or at different temperatures. It may be further useful to observe the device with a microscope in the visible and/or infrared wavelengths (silicon is transparent to wavelengths greater than 1.1 micrometers) while the device is undergoing testing (see section below on observation modes).

Later in the development program, the MEMS device yields in larger quantities. The characterization test station now needs to be able to test many devices at once. At this point, test speed is prioritized over flexibility. For example, a qualification test may require 100 devices to be tested. To test these many devices quickly, the characterization test hardware will need to be scaled to test 100 devices, or some substantial fraction of that, at one time. Smaller test volumes would require the development team to load, test, unload, and load the next batch in order to test all 100 devices. This is laborious and time consuming, and it requires the development team to spend time on running the test rather than on interpreting test results.

Scaling the test volume to evaluate multiple devices simultaneously is nontrivial. At this point of the development program, the single test station will have been iterated and debugged. For the higher volume test station, the single test station's designs should be reused as much as possible, so that the development focus can be on the scaling of test volume, and not on redeveloping capabilities already established in the single test station. To plan for this, it is useful to design the single test station modularly so that the components, such as input stimulus and electronics, can be independently scaled.

As test capabilities grow, the early test station should not be discarded when the later, higher test volume test station becomes available. Rather, the early test station can be reserved, and perhaps upgraded, to do more in-depth failure analysis of individual parts.

The logical endpoint of test capability evolution is the equipment to be used in manufacture test, at which point, thousands or millions of devices will be tested at high throughput. See Chap. 20 for discussion of manufacturing test.

Information Quality vs. Test Speed and Volume

There is a trade-off between information quality and test speed. It is often useful to have at least two types of test stations: one that does a more complete, higher quality, but slower evaluation, and one that does a faster, but less thorough evaluation. For example, it would be useful to have a test station that tests one device thoroughly, and another station that tests 100 devices rapidly. The 100-device test apparatus can quickly scan multiple devices for failures. The single device apparatus would then test the failed devices more thoroughly to conduct failure analysis.

As an example, a pressure sensor development team is searching for a particular device failure, which occurs in a few percentage of the devices. The team would use the high-volume test station to scan 100 devices quickly by sampling at only a few pressures. One part is found that fails to meet the specification. That part would then be taken to the slower, more complete test station that, for example, sweeps pressure in 0.1 atmosphere steps through the pressure range and also sweeps temperature and voltage, while allowing optical observation of the part, to learn more about this part and why it is failing.

Testing During vs. Before and After Exposure to Conditions

In addition to the characterization testing, test capability to impose operating conditions and reliability conditions will need to be established. Examples include equipment to expose the MEMS product to conditions such as mechanical shock, humidity, and the HAST test.

Ideally, in these operating condition and reliability tests, a device's performance is monitored before, during, and after exposure to the environmental stimuli. But, monitoring performance during the exposure can be a daunting and expensive engineering challenge. Continuing the TPMS pressure sensor example, if one were to monitor performance of the pressure sensor during a HAST test, the test machinery would need to have a means to modulate the input pressure and readout the sensors while the devices are exposed to the HAST conditions. The test machinery, including the electrical lines, connectors, boards, fixtures, data acquisition hardware, and many other components, must therefore be able to tolerate the HAST conditions. Since HAST is intended to stress products, it will also stress the test machinery. As a result, developing test monitoring capability for HAST can be a major engineering undertaking.

Because developing a test apparatus that can monitor device performance during these conditions may be difficult, it is more common to evaluate the device before and after, but not during, the treatment and compare before and after results to see if there are any changes.

However, before and after test performance monitoring does not allow observability during the exposure, which is useful for failure analysis. If, for example, the pressure sensor does not pass HAST, then all that is known about the failure is that the sensor did not pass. There is no information on how it failed. But if the sensor had been monitored during the test, then it could be known when and how the it failed. Did failure occur early in the test? Late in the test? Did the sensor suddenly stop working or did it slowly degrade? If it suddenly stopped, maybe a fracture or crack developed in the sensor membrane. If it slowly degraded, then maybe there was a slow leak in the vacuum reference cavity. Electrical monitoring may reveal telltale signs, such as a short circuit, that may indicate water intrusion or an open that may indicate galvanic corrosion at a contact. All of these are important clues that help to focus the failure analysis investigation.

A balanced approach is to first develop the before/after device test capability. This is faster and more cost-effective, and it allows the development team to evaluate whether the capability to monitor performance during HAST exposure is needed. If monitoring during exposure is needed, a compromise capability is the ability to monitor the output of the devices during test, such as the bridge impedance, but not apply stimuli, such as pressure. This can give useful information such as whether the failure occurred early or late in the test, and whether, for example, the failure was sudden or underwent a slow degradation.

Developing for In-Field Test Capability

Developing the test capability for in-field testing also requires engineering resources and should be planned for in the product development program. Identifying these requirements early in product development can prevent re-engineering of the test equipment. For example, a portable, ruggedized, battery-powered version of the test board that can readout the MEMS and log the data could be developed for multiple uses, include in-field testing, drop testing, and other situations where wireless, battery-based operation is useful or convenient.

For a tire-pressure monitoring device, a relevant example is testing the device while it is in the tire of a moving car. Wires cannot be connected to the device under test, because of the tire's continuous rotation. Instead, wireless, battery-operated boards that can handle the tire's conditions (the rotation, road bumps, etc.) can be developed. These boards could also serve a dual purpose of monitoring the device during mechanical shock testing. Recognizing this testing need early in the development allows the boards to be designed so that they also would meet the requirements of both tests.

If the development program plans to develop a custom ASIC for the MEMS, it would be useful if the ASIC can also be programmed to function as a datalogger. Then the ASIC can serve as the datalogger on a battery-operated test board. If a custom ASIC will not be developed, then a potential substitute is to employ an off-the-shelf microcontroller. Chapter 13 discusses microcontrollers in greater detail.

Best Practices for Enabling Failure Analysis

Once a failure mode has been identified, the next step is to understand its mechanism. One general challenge is the frequency of a failure mode's occurence. Part per 10 failure modes are easy to observe because they are so frequent, but part per million (ppm) failure modes are difficult to observe, because they are so rare. It may be necessary to test several million devices in order to gather a small population of failed parts to conduct an analysis.

Furthermore, typically, the rarer the failure mode, the more esoteric the failure mechanism and the more difficult it is to replicate and understand. For example, a device may fail due to temperature exposure at a part per 100 rate, and fails due to humidity exposure also at a part per 100 rate. Both of these failures are observed, analyzed, understood, and then fixed in the device's design and/or process. A new iteration of the device now passes separate exposure to heat and exposure to humidity tests. However, it subsequently fails under simultaneous exposure to both heat and humidity at a part per 1000 rate.

In MEMS products, it is possible to have a failure mechanism which requires two (or more!) conditions to simultaneously occur. Continuing the above example, a warm temperature condition expands small imperfection or hole in a package which allows water vapor to enter. If only the temperature condition were present, then there would be no humidity to enter the package. Conversely, if only the humidity condition were present, then the hole would remain closed due to lack of heat. This failure mode only occurs when both conditions are simultaneously present.

The complexity and rarity of failure modes increases as more conditions are added. Continuing the above example, an updated design and process is implemented and addresses the warm, humid failure mode described above. Now, the device passes exposure to warm, humid conditions, but the design fails under the combination of warm, humid, and temperature cycling conditions. Water vapor transported inside the MEMS during a warm cycle condenses to water droplets during a cold cycle causing a different failure. With this discovery, the design and process is again updated and tested again.

Identifying the nature of a failure can be an arduous task. Failure analysis does not follow any algorithm other than the scientific method of forming a hypothesis and designing experiments to prove or disprove the hypothesis. To test and understand a failure mode, patterns in the device behavior must be observed. To do this, many devices must be poked and prodded with various stimuli and measured to see how they react.

There are some general strategies for identifying failures in MEMS devices.[2, 3] One is to have a good understanding of the method of Design of Experiments, and there are many good books that discuss this topic.[4] Other specific strategies helpful for de-bugging MEMS devices are discussed below.

Localize the Problem

One general strategy is to localize or isolate the problem by component and domain. When isolating by component, ask the question: is the problem in the MEMS die, the electronics, in the package, or in the test system? When isolating by domain, ask: is the problem mechanical, electrical, thermal, optical, chemical, or other domain? Localizing the problem, to say, an electrical problem on the MEMS die or a mechanical problem in the package greatly helps to focus the failure analysis efforts.

To localize the problem, it is often useful to have variant devices that isolate each component and domain. For example, a variant pressure sensor might be one that consists only of its electrical components by removing the mechanical component, i.e., the pressure-sensitive membrane. Another variant may be pressure sensors with different bridge resistances to modulate electrical behavior. Another may be pressure sensors with membranes of difference sizes to vary their sensitivities to pressure. Yet another is a plain dummy die with no MEMS device and only the electrical bond pads for wirebonding. To have these variant devices available, they must be planned, so that they can be included in the mask layout.

To illustrate the use of device variants for problem localization, we can use an example where an electrical open in the package occurs under simultaneous humidity and temperature conditions. A variant of the MEMS product could be built with a dummy die substituted for the MEMS, i.e., a MEMS die with no electrical or mechanical functionality, only wirebond pads to electrically connect to the package. If this variant continues to demonstrate the electrical open behavior, then it rules out the problem being in the MEMS die, and localizes the problem to either the electronics or the package.

For the electronics, failure localization can be done by interchanging the electronics. This requires a MEMS device that can be read by different electronics, such as by both a board and benchtop instruments. If, for example, the MEMS device malfunctions when connected with a readout board, but not with the benchtop instruments, this isolates the problem to the board. Being able to do this substitution requires some forethought. For example, if the device under test is soldered to the board, then substitution is difficult. If, instead, sockets were chosen for electrical connection, the device under test can be moved easily from the board to the benchtop instruments. There is more discussion on how to use interchangeable proxies for the MEMS, electronics, and package in Chap. 13.

Continuing the above example where the dummy MEMS die suffers electrical opens under simultaneous humidity and temperature conditions, the development team substitutes the board electronics for more mature and debugged bench electronics. The product is tested again and it still demonstrates the electrical open behavior. At this point, the development team's search for the failure mode has been narrowed to an electrical problem in the package. The development team forms several hypotheses. One hypothesis is that galvanic corrosion causes a bondpad to degrade and lose electrical contact to the wirebond. At this point, the team has localized the problem so that microscope inspection may confirm the hypothesis. The team observes the bondpads under microscopy and is able to confirm bondpad corrosion.

The strategy of localization includes the test system. Typically, the test system is undergoing parallel development along with the MEMS product. As a result, it also needs to go through its own design-build-test iterations. In the example cited earlier, with the building HVAC cycling on and off causing intermittent failure, it was assumed that the problem was in the MEMS device. If the test system had had a way to localize the problem between the test system and the device under test (DUT), then weeks might have been saved. It is therefore useful for a test system to have a method of self-test and calibration. Self-test could be implemented, for example, by having a reference sensor in the test system to demonstrate that it is functioning as expected.

Observation Modes

One of the challenges with MEMS, or any type of microdevice, is that problems are hard to detect because the devices are small and may contain buried features that make it hard to observe a problem directly.

The two most common modes of observation for a microdevice are external and in situ. External observation includes methods such as optical microscopy, infrared microscopy (see Fig 14.3) [5], white light interferometry, acoustic microscopy, and electron microscopy. External observation tends to be easier since the device needs only to be placed in the instrument and observed.

In situ observation uses the structures in the MEMS device for observation. An example is the capacitive comb fingers or plates within MEMS accelerometers and

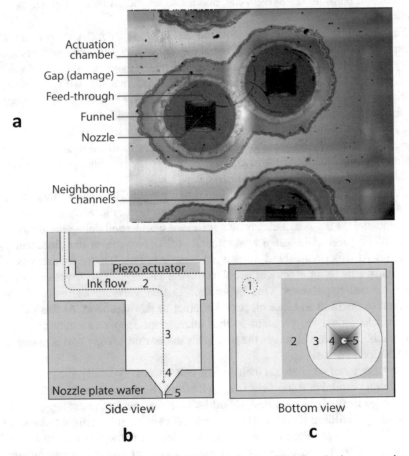

Fig. 14.3 (**a**) Nozzles of a silicon MEMS inkjet printhead under infrared microscopy showing damage. Silicon is transparent to infrared radiation at wavelengths longer than 1.1 um. One can see the nozzle at the silicon surface, as well as the funnel, feedthrough, and actuation chamber under the silicon surface. (**b**) Sideview of the printhead. (**c**) Bottom view illustration of the printhead. (Reprinted with permission from Lohse et al. 2011[5])

gyroscopes. These can be very precise, for example, measuring femtofarad magnitude changes, which can correspond to displacements as small as a nanometer. In addition to being precise, in situ observations are often electrical in nature and their interrogation can be automated by data acquisition hardware. This allows large sets of data to be collected and analyzed.

Continuing this example, an accelerometer's capacitive comb fingers can also be used to measure electrical shorts in the device and capacitance-voltage (C-V) curves, which are useful in probing the accelerometer's proof mass's spring constant as well as the gap between the moving capacitive electrodes. Finally, an accelerometer's capacitive output can be used to obtain a mechanical spectrum of the proof mass motion, which can give information about the proof mass's dynamics, such as its resonant frequencies and modes.

All of these capabilities are useful in MEMS product development, but to reap their benefits, they require planning. Mechanical vibration spectral response, for example, is a challenging measurement, requiring high-speed data acquisition of small capacitive signals. It is helpful if the device itself has features to facilitate acquisition of a high-resolution mechanical spectra, such as Faraday shielding, to protect the small capacitive signals.

External and in situ modes of observation are complementary, and simultaneous observation from both modes is very useful. For example, if one can monitor the motion of the proof mass of an accelerometer through both external and in situ methods, then one can, for example, correlate capacitive sensor readings with externally observed proof mass states. Suppose a particular capacitance value recurs over and over again. Infrared optical observation of the proof mass may show that this particular capacitance measurement occurs when the proof mass is immobilized due to stiction at a particular position.

Using Test Structures

Modes of observability are enhanced by using test structures. Test structures are specialized structures that have been designed specifically to characterize a particular phenomenon. Some examples of phenomena that test structures may be used to investigate are:

- Electrical properties, such as contact resistance, material resistivity, and parasitic capacitance.
- Material properties, such as internal stress and failure stresses.
- Fabrication issues, such as misalignment effects and etching artifacts.

For example, if there is a risk that thin film stresses in a device may cause issues, then a test structure to characterize these stresses can be included in the wafer layout. There are a number of thin film stress test structures, including Gückel rings, cantilevered beams, and verniers with pointers [6]. The most appropriate test structure will depend on the expected stresses (e.g., tensile vs. compressive), the method of observation, the fabrication process, and other factors.

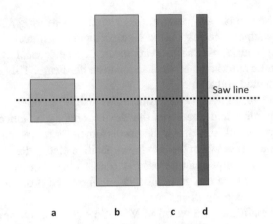

Fig. 14.4 Test structure for SEM observation inside a cavity. (**a**) is the cavity inside which observation is desired. This cavity is small in area and therefore difficult to get a saw line to intersect it. (**b**) A simple test structure is a lengthened version of the microcavity to facilitate intersection. Moreover, several test structures (**c**) and (**d**) intended for SEM observation can be arranged together so that a single scribe and break or saw pass allows observation of all the cavity cross-sections and also allows easy navigation from structure to structure while observing in the SEM

In another example, suppose the development team would like to see inside a micro-cavity (see Fig 14.4). It is expensive and time consuming to precisely cut open a small cavity. Instead, a useful test structure is a longer version of the cavity so that an imprecise scribe and break of the test structure will open it and allow interior observation (scribe and break is discussed in the singulation section of Chap. 12). Furthermore, if you anticipate needing to take multiple SEM images of different test structures, they can be arranged in a line so that a single break or dicing saw pass allows observation of all the desired components. This will also allow easy navigation from test structure to test structure under microscopy.

While many different test structures are possible, they should be selected for their compatibility with the fabrication process for the device. A different fabrication process may alter the test structure so that it is no longer similar to the device, and therefore not useful for the device's characterization. Moreover, a separate fabrication run solely for a test structure is usually cost prohibitive.

In some cases, the presence of test structures can also increase fabrication risk and decrease device yield. The risk of any test structures needs to be carefully weighed against their benefit. A set of cantilevered beams, for example, may be an excellent test structure for characterizing residual stress of a thin film. But the beams may break during a spin-rinse-dry step in the process, thereby creating debris that could cause loss of the entire wafer.

You should budget to design and measure each test structure and device variant. Device variants and test structures are devices in their own right and therefore require thoughtful design and layout. After fabrication, time must be spent measuring and characterizing them. There will usually be more test structures and device variants to include in a mask layout than budget and time available to actually test them. Hence, prioritize and limit the test structures to what is justifiable and practical.

Summary

Testing is an often overlooked but important part of the MEMS development program. It is a critical capability that offers significant opportunities to accelerate the MEMS development, but requires planning and investment. The requirements will define the testing program, and therefore need to be carefully considered. Development of the test capabilities is a significant undertaking that will require design-build-test iterations to establish capability that can allow the development team to focus on the MEMS development and not on debugging the test stations. In addition, the test program will intersect with aspects of the MEMS, electronics, and package development programs. These intersections offer opportunities to identify and implement strategic test solutions that can meet the needs of the entire product development program.

References

1. (2015) Highly accelerated temperature and humidity stress test JESD22-A110E. JEDEC. https://www.jedec.org/standards-documents/docs/jesd-22-a110c.
2. Hartzell AL, DaSilva MG, Shea HR (2011) MEMS Reliability. Springer, New York
3. Costello S, Desmulliez MPY (2013) Hermeticity Testing of MEMS and Microelectronic Packages. Artech House, Boston
4. Montgomery D (2013) Introduction to statistical quality control. Wiley, Hoboken, NJ
5. Lohse D et al (2011) Infrared imaging and acoustic sizing of a bubble inside a micro-electro-mechanical system piezo ink channel. Journal of applied physics 110(3):034503
6. Masters N, de Boer M, Jensen B et al (2001) Side-by-side comparison of passive MEMS strain test structures under residual compression. Mechanical Properties of Structural Films. ASTM International

Chapter 15
Risk Mitigation Strategies for Prototype Fabrication

Building a prototype is a critical, essential step towards validating a new MEMS design. It is an exciting undertaking to see any design realized on a wafer. It can also, however, be a frustrating, time-consuming, and very expensive experience. No one wants to spend weeks or months processing wafers only to find that the prototype device will not function; or worse yet, that process conditions caused the MEMS devices to break mid-process for some mysterious reason.

In this chapter, we introduce a few simple strategies which will help to increase the probability of prototype success and to lower the risk that prototype wafers will end up tossed in the recycling bin, not to mention saving time and money. If you are an industry-experienced MEMS process engineer, you might skip ahead to Chap. 16. If you have only ever worked in academic or research fabs, this chapter is for you.

Knowing Your Organization's Tolerance for Risk

It is common for engineers to want to attempt a fully featured prototype as soon as possible. After all, that is what will impress investors and executives. However, due to the large number of variables in even a short MEMS process flow, going from a new mask set to a fully fabricated prototype after a single run is only realistically possible in a few specific cases, such as:

- Simpler MEMS devices where process risk is low, such as a microfluidic device with easy process tolerances or a device with only a few mask levels (<4).
- Designs which are based on a well-developed MEMS technology that is documented and well understood, such as a pressure sensor.
- Devices that will be built on a foundry platform process, following the foundry's design rules.
- MEMS designs that can be comprehensively modeled with very few design, material, or process unknowns.

© The Author(s), under exclusive license to Springer Nature Switzerland AG 2021
A. M. Fitzgerald et al., *MEMS Product Development*, Microsystems
and Nanosystems, https://doi.org/10.1007/978-3-030-61709-7_15

In the case of most MEMS prototyping efforts, you should prepare yourself, your colleagues, and your investors or executives to expect to process *multiple* proof of concept and prototype wafer lots in order to reach the stage where the prototype truly represents the final device, your intended design.

An important first step to developing a solid prototyping project plan is to assess your organization's culture and attitude towards risk. If you do not already know how your executives and investors feel about risk and failure, you must find out now. For example, if your company's decision makers are highly failure- or risk-averse, then a prototyping project plan must be organized to enable slow, steady, progress with a series of small successes, so that they will feel reassured that you are making good progress, and so that they will not terminate the project in case of an early failure before there are any useful results. If your company's executives or investors are undeterred by short-term failures and patient, then your project plan will be able to tolerate more risk and you would not have to fear your project being terminated when the first wafer breaks or fails to meet a design specification.

We pass no judgments about a company's attitude towards risk of project failure; we know that company cultures range from carefree to highly risk averse, no matter the funding situation. Which risk mitigation steps will provide the most effective approach is situational, as no one path is right for everybody and every company. You can and should make different choices about where to expend effort and money depending on your company's unique circumstances.

Analyze Process Risks and Explore Trade-Offs

Cost, time, risk—similar to the adage about faster, cheaper, and better—you typically can only reduce two at the expense of the third. Knowing the balance you need to strike among cost, risk, and time, in order to keep your executives or investors happy, choose two factors to optimize in planning a fabrication project and its resources.

Once you have that context, performing a process risk analysis before beginning prototype fabrication can yield valuable insights on how and where to expend resources in order to improve chances of yielding successful prototypes.

There are many well-established failure analysis methods, such as FMEA, which provide useful templates. These will help you to identify areas of highest process risk, specific to your device design, that will deserve special risk mitigation techniques. Descriptions of these failure analysis methods are beyond the scope of this book; however, we refer you to the cited references for more information [1–3].

Many of the common failure analysis methods were originally developed for traditional manufacturing methods and environments; however, they are readily adaptable to MEMS manufacturing conditions. We include a very rudimentary example of a FMEA-style process risk analysis for a SOI wafer-based pressure sensor process (Table 15.1). We will also revisit this example at the end of the chapter in order to help illustrate various risk mitigation methods.

Table 15.1 Partial/simple FMEA-style risk analysis example for a MEMS pressure sensor. The acronym COC stands for "Certificate of Compliance"

Process module	Process risk	Probability	Impact	Risk mitigation action
Starting SOI wafer	Dopant level in device layer silicon	Low	Medium	Obtain COC, perform SRP analysis after implant
	Thickness of device layer silicon	Low	High	Obtain COC, perform cross section SEM after DRIE to inspect membrane
Ion implant	Photoresist exposure for defining resistor implant	Medium	High	SEM inspection of photoresist CDs and sidewall profiles
	Dopant dose and energy incorrect	Low	High	COC from implant vendor
	Dopant dose and anneal incorrect	Medium	High	Model implant dose and anneal, cross-check with post anneal SRP data
	Resistor value out of spec	Low/ medium	Medium	Electrical probe test after metallization, before DRIE, at a few locations on wafer
DRIE of handle wafer silicon to form membrane	Back side alignment to front side features	Medium	High	IR alignment check, cross-section SEM post etch with doped region highlighted
	Sidewall etch angle	Medium	High	Cross-section SEM post etch with doped region highlighted

The aim here is to create a robust fabrication plan by taking the time to think through potential risk items and how to address them; you do not need to figure out all the details just yet.

In MEMS fabrication, no matter how experienced you are, chances are you will not be able to anticipate and protect against all possible process risks. However, identifying most of them will definitely get you on the path towards success. For any MEMS device, there are usually many possible ways to process it, so this risk assessment exercise will also be a great way to compare and select among candidate process flows. It may also identify, for example, the need to purchase extra SOI wafers, knowing now that some may need to be sacrificed mid-process for inspection, or to locate a vendor that can do spreading resistance profile (SRP) analysis of an ion implanted region. We promise that you will not regret taking time to do a process risk assessment even if all of your process modules are known to have low risks!

Process Development Risk Mitigation Strategies

Once the process risk analysis is complete, turn your attention to determining how to best address each identified risk. Chapter 10 outlined the best practices for reducing risk during the design/modeling phase. In this chapter, we provide several

Table 15.2 How different process risk mitigation techniques address project cost, time, or risk

Risk mitigation technique	Cost	Time	Risk implications
Short loop	Inexpensive way to discover helpful process information early on, without having to execute a full process flow	Adds some time at the beginning of the fabrication plan in order to conduct experiments	Mitigates risks that are known at the start of the development. Data from short loops will helpfully inform downstream design and process choices
Staging, split lots, or look-ahead wafers	Adds some cost because more wafers must be fabricated early in the process to enable some to later be helpfully split off from the main batch	Avoids having to start over from scratch when a process fails at step N, because wafers following behind at step $N - 1$ will be available to continue	Only the look-ahead wafer or split lot would be lost, not all wafers
Parallel lot processing	More expensive: cost multiplies by the number of wafer lots being processed in parallel	Fastest way to gather information on multiple design or process variants	When combined with short loops or staging, this is a powerful method to reduce overall development risk and time

strategies that may be used to manage risk during MEMS prototype fabrication, whether in the proof of concept, prototype or foundry feasibility stages of development. We describe three common categories of tactics:

- Short loops (a small subset of the full process flow).
- Staging wafers, split lots, and look-ahead wafers.
- Parallel lot processing.

Short loops help to explore and manage known risks in a short period of time and for a smaller budget. Staging wafers, splitting the lot, or using look-ahead wafers helps you to better deal with and recover quickly from any surprises that might arise along the way. Parallel processing requires a larger budget but enables you to quickly explore and select from several possible development pathways. Some of the characteristics of each of these three techniques are summarized in Table 15.2 and explained in more detail below. Keep in mind that these three are not mutually exclusive.

If you take time to consider how to best deploy these different process risk mitigation tactics in your process development, in the context of your company's attitude towards risk, you will be able to figure out where to spend money, where to take more time, and where to take chances.

Short loop Process Experiments

Short loops are a powerful investigative tool which can be used to reduce process risk and resolve uncertainties in early prototyping, in the feasibility stage when first transferring to foundry, or even when making design changes in an established device.

For short loops to be most effective, they need to be short! You should not design a short loop that requires 50% of the full process's steps. Ideally, you should be able to execute the short loop in 1–2 weeks. The goal is to simplify a portion of the process enough to focus down on key issues or unknowns, but not so much that important contributing conditions from preceding steps (e.g., material interfaces, surface roughness, thermal budget, topography) would be ignored. If you think you might need to execute 50% of the process flow before reaching the process conditions you hope to investigate, you should consider staging wafers instead of using short loops.

When selecting which process modules or steps to short loop, focus on:

- Identifying steps that are new to you or the foundry.
- Steps where you might need data on process tolerances to inform the MEMS device's design or model.
- Any step critical to the device's function.
- A step from which there is no recovery or rework option, such as a wafer bond or deep silicon etch.
- Any process high-risk items identified in your risk assessment.

Some examples of process modules and results that benefit from exploration in short loops include: use of non-standard photoresists, DRIE performance on a new pattern, thin film stresses and morphology, wafer bonding, alignment tolerances, etch selectivity, and sacrificial material release.

If designed thoughtfully, multiple short loops could be run in parallel, thus minimizing time needed before the full build process could begin (as shown in Fig. 15.1). For example, you might execute multiple parallel polysilicon depositions on test wafers in order to discover the ideal process temperature which yields the grain morphology best suited for your device. You may also parallelize short loops that address different points within a lengthy process flow, such as examining polysilicon grain structure (an early step) and deep reactive ion etch (DRIE) performance (a later step in the process flow).

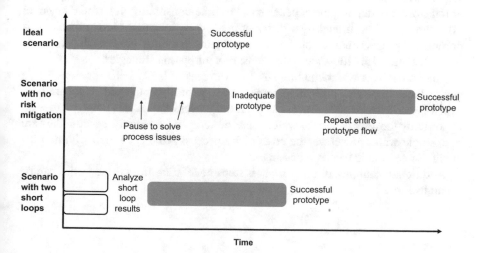

Fig. 15.1 How short loops can be used to decrease total process time

It is also useful to conduct short loops while a batch of wafers is being run through the full process flow. For example, you might begin running a process flow that has simple early steps, and then while the wafers are running, in parallel, perform short loops to investigate tricky later steps. By the time the process wafers reach the tricky step, you will have the information from the short loop needed to successfully execute the tricky step.

Short loops are also very useful for surfacing additional IP or trade secret information, such as unique process conditions or features which might be needed to elicit better performance from your device design. Discovering this information sooner allows you to file patent applications earlier or to establish which process steps need to be siloed and kept confidential.

Short loops are also an effective tool to increase confidence in a process and a design at all levels in your organization—for the people processing the wafers, as well as for the people funding the development.

Staging, Split Lots, and Look-Ahead Wafers

Another technique for reducing process risk is to start more wafers than you need and then to hold, or "stage," wafers at critical points in the process flow. In order to identify the staging points, you should refer to your process risk table that identified major risks in the process. You can then determine if the risk point is best managed by staging wafers, running split lots from that point or running one or two look-ahead wafers.

A look-ahead wafer is just as it sounds: send one wafer through the next process step ahead and then once that step has been successfully confirmed, put the rest of the wafers through that step. This tactic works best when you will know the results of that one step immediately, confirming that the process step is safe. If the process step is such that you cannot know if it has been successful until several steps later, or if the process step is quite expensive and/or time-consuming, you would be better off running wafers through two different process options (split lots) or holding a portion of your wafers at the completion of the previous process step (staging wafers). Figure 15.2 illustrates these three risk mitigation strategies.

Silicon DRIE or wafer bonding are two examples of MEMS process steps where any of these three risk mitigation steps can be helpfully employed. They are both expensive and irreversible process steps; if the process fails at this point, there is no opportunity for rework and the wafers must be scrapped. Secondly, because they are both single-wafer processes, the costs of the process scale per wafer, with no possibility of economic batch processing.

Additional examples of other process steps benefitting from these tactics can be found in Table 15.3.

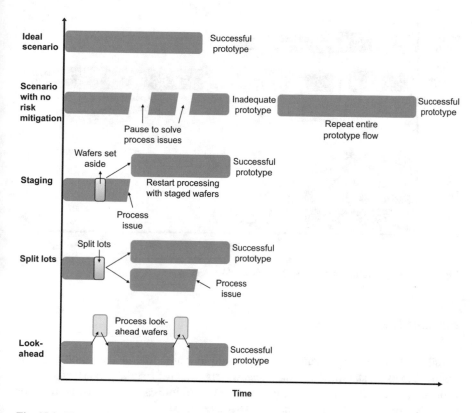

Fig. 15.2 How staging, split lots, and look-ahead wafers can be executed

Table 15.3 Typical higher-risk MEMS process steps which warrant use of risk mitigation strategies

Process step	Comment
Deep reactive ion etch (DRIE)	Irreversible and expensive single wafer process
Deep anisotropic wet etch, such as KOH	Long process that requires good masking and crystal plane alignment for best results
Wafer bond	Irreversible, single wafer process which usually comes late in the process flow, creating high risk for valuable wafer loss
Lithography over large topography	Good resolution of features will depend on topography and exposure method
Hermetic cavity	Many process variables influence the hermeticity of a cavity
Sacrificial material etch	Always requires tuning to execute cleanly without damage to adjacent materials
Polycrystalline material deposition	Multivariable deposition conditions require tuning to achieve desired material properties.
PECVD material deposition	Multivariable deposition conditions require tuning to achieve desired material properties.
Timed etch processes	Dependence on time for etch stop makes accurate process control and holding dimension tolerances difficult

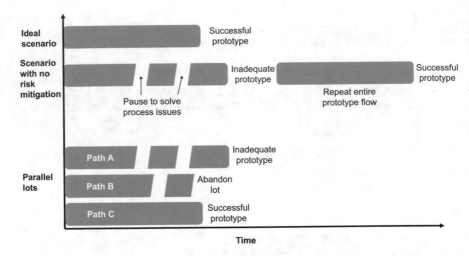

Fig. 15.3 Using parallel lot processing for faster discovery

Parallel Lot Processing

Parallel lot processing means running multiple wafer lots at the same time, as opposed to using staging and look-ahead wafers, which are serial processes. Parallel processing may be employed to rapidly explore multiple candidate process flows or mask designs, or both at the same time (See Fig. 15.3).

Parallel wafer processing can dramatically speed up discovery and product development, in exchange for much higher cost. Unless you are in the rare position of having a generous or unlimited development budget (lucky you!), it needs to be used carefully. When there is significant time pressure on a product development, exploring multiple process paths simultaneously through parallel lots will save a great deal of time. However, not all paths may yield actionable results, and discoveries from one failed path may not necessarily be made in time to inform another ongoing process path.

Cost-Effective Planning

When deploying these techniques, you will need to thoughtfully manage the number of wafers, process wafers that will never be completed, and diligently track where all the wafers are in the process flow at any given time.

To determine the total number of wafers needed for staging or split lot, you must first determine the number of fully completed wafers you hope to have at the end of the process. Then work backwards through the process in order to identify the stage and splits points and sum up the number of wafers that will be needed at the start of the process.

For example, when planning a proof-of-concept or early prototype process, we recommend starting a minimum of three wafers in a group in order to have a high probability of at least one wafer surviving the entire process flow. Then if you would like to have one staging point in the process where you want to set wafers aside, you will need to start at least six wafers.

When planning for a split lot, every process flow split group should each have at least three wafers. For a process with one split, you will need to start a total of six wafers. If you combine both of these examples, one stage (setting aside three wafers) followed by one split (three wafers in each lot), then you will need to start nine wafers.

While parallel processing two lots would simply double the number of start wafers, in order to maintain a minimum lot size, staging wafers and splitting process flows could increase the number of wafers you would need well beyond a factor of two. Every added wafer lot increases the cost of the process run and the complexity of managing it, so there is a practical upper limit on how many stage and split points can be utilized. Again, your risk mitigation table can help you to strategize the use of split lots and to identify when they would best be deployed.

Do not take a combinatoric approach and try to process every combination in order to identify an optimal process flow. This can be a useful thought exercise but it is not an efficient processing plan; if you perceive that much risk in the process, a better strategy would be to reconsider the design or to employ modeling to identify a safer process flow (see Chaps. 10 and 11).

Ultimately, the cost of carrying the extra wafers through a process to enable staging and splits depends on the expense and type of process steps (batch vs. single wafer processing). Look-ahead wafers may seem the most cost effective; however, keep in mind that you will be repeating the process module with the rest of the wafers in the lot. If you were to use a look-ahead wafer on every single step of the process, then in total, you would be executing the full process flow twice: once for the look-ahead wafer and once for the remaining wafers in the lot. Those processing costs will add up very quickly over a long process flow.

After Spending the Money and Time, Be Sure to Harvest the Data

Because there is extra cost, time, and effort to deploy these process risk mitigation steps, it is important that their full value be captured by making sure all available data is collected, documented, and analyzed from all of the wafers, not just the successful ones.

You may end up with partially processed wafers that were left behind and never finished. Each remaining wafer may have a different processing history. If you make the effort to carefully document the history of each wafer's or lot's particular path, that data can be very valuable and used for informing future process design, process execution, foundry transfers, training new personnel and for analyzing test results. The wafers themselves may also be used in follow-on short loop experiments, or for other aspects of development such as back-end process experiments and packaging development. This value can only be captured if you are diligent with documenting each wafer's history and then harvesting its data.

Example: Using Fabrication Risk Mitigation Strategies

All of the above techniques are useful to reduce risk of failure during process flow development. Although some may increase the time to the completion of a given wafer process run, they will greatly increase the probability of having useful and working devices at the end of that run. It will be worth the wait—after all, what would be the point of a fast fabrication run that yields nothing useful? Ultimately, consistently using these techniques will reduce the overall time to develop a process that is ready for production. Reducing the *total* elapsed duration of product development is what matters most.

Figures 15.1–15.3 illustrated how one might implement these techniques. In Table 15.4, we provide some examples of using them in an SOI wafer-based pressure sensor process flow.

Process Risk Mitigation in a Production Fab

The techniques described in this chapter may also be used when transferring an advanced prototype process to a production fab or foundry. In the production setting, it is generally best to avoid split lots due to the extra cost and the effort required

Table 15.4 Example of deploying risk mitigation strategies in a SOI wafer-based MEMS pressure sensor process flow

Process module	Step	Strategy	Manages risk
SOI wafer	Inspect and mark wafers; divide into two lots	Parallel processing: Two SOI wafer lots with different device layer thicknesses	Pressure sensor sensitivity
Ion implant	Implant and anneal	Split: Two lots with one dose, implant energy and two anneal recipes	Modeled vs. actual implant concentration and annealed profile
Metal	Deposition	Look ahead: Deposit and pattern metal, electrical test	Ohmic contacts formed
Form membrane	Back side lithography for DRIE	Stage: Hold X number of wafers after lithography, before etch	DRIE alignment and etch profile

to track the wafers in that setting. It is not what the manufacturing infrastructure is set up to do. If a foundry does allow split lots, there will likely be a minimum lot size of 10 wafers per split in order to maintain an adequate volume of wafers throughout the entire process. When added up over multiple splits in a process flow, the foundry may end up needing to start 50 or more wafers. This number of wafers is quite expensive to carry during the feasibility stage at a foundry, when process costs will still be quite high.

Development and research fabs are much better suited for processing smaller and more frequent split lots; therefore, the most appropriate time to do process splits is in the advance prototyping stage, not at the foundry. Please refer to Chap. 17 for more information on the trade-offs between prototyping at a development fab versus a production fab, and how to assess readiness for transfer to a production fab.

Summary

Risk is an inherent part of developing any new technology. Understanding your company's tolerance for risk is an important first step in evaluating your options to reduce it. Before processing your first MEMS wafer, look for ways you can lower risk within the design and process flow. This is the most effective way to reduce risk at the beginning of development.

Short loops, staging wafers, split lots, look-ahead wafers, and parallel processing are all effective ways of reducing risk during MEMS wafer processing and they can be used throughout all stages of development. Your available budget will play an important factor in deploying these tactics. Proper fabrication planning enables you to best control the number of wafers starts, and therefore your expenses, throughout development.

References

1. Carlson C (2012) Effective FMEAs: achieving safe, reliable, and economical products and processes using failure mode and effects analysis. John Wiley & Sons, Hoboken, NJ
2. Stamatis DH (2003) Failure mode and effect analysis: FMEA from theory to execution, 2nd edn. ASQ Quality Press, Milwaukee
3. McDermott RE, Mikulak RJ, Beauregard MR (2009) The basics of FMEA, 2nd edn. CRC Press, New York

Chapter 16
Documenting MEMS Product Technology for Transfer to Manufacturing

For a developing MEMS product to make the transition from an R&D facility to a manufacturing fab or foundry, the product technology must be passed from a team having intimate knowledge and history of it, to a team that perhaps has never encountered the product technology before. It is a significant communication task that deserves attention.

Documenting a fast paced and complex MEMS development and periodically reviewing it for completeness, especially paths that were ultimately not successful, is every engineer's least favorite task and can often be considered a "nice to have." However, neglecting documentation could end up costing your company in the long run.

In addition to creating a technical dossier for enabling transfer of processes to foundries or other production facilities, development documentation can help to establish and preserve your company's IP and know-how, to avoid repeating work already done, and to help new team members integrate quickly. It is also important to create a written history of your MEMS development because many subtle yet important details can easily be lost in time, due to departing employees and fading memories. What seems so embedded in your brain right now, you might be trying to reconstruct a year later when revisiting a design decision.

What Needs to Be Documented and Why

There is currently no standard for documenting MEMS technology for transfer to manufacturing. Established companies will have documentation methodologies and practices already in place, especially if they are operating under ISO-9001 or other quality standards. If this is the case at your company, the contents of this chapter will help you to fine-tune your existing documentation practices for the specific needs of transferring a process to a MEMS foundry.

© The Author(s), under exclusive license to Springer Nature Switzerland AG 2021
A. M. Fitzgerald et al., *MEMS Product Development*, Microsystems and Nanosystems, https://doi.org/10.1007/978-3-030-61709-7_16

We have observed that start-up companies tend to be most neglectful of documentation, due to factors such as inexperience of the team or having a narrow focus on fast-paced work. The contents of this chapter are specifically aimed at guiding start-ups, as well as companies new to MEMS technology, on how best to document their hard-won development knowledge.

Starting a documentation protocol early in product development will enable you to build a knowledge database in an organized manner. Good documentation does not mean everything must be written down; if you were to document absolutely *everything,* you and your colleagues would be smothered in details and have trouble sorting through them to find essential information. The goal is to proceed with intent to make sure that you have captured the significant information needed for manufacturing and also why certain choices were made.

Keep in mind that a device or process technology is not fully under your control unless all the "knobs" that make it work have been identified and all of the knob settings that make it perform well are understood. The point of technology development is to discover and explore all the knobs and their settings. Once those are in control *and documented*, it is possible to mass produce a product.

The following questions can help to efficiently focus documentation efforts:

- What is new or special about your design/process/product?

  ○ Document how your innovation or process departs from standard practice.
  ○ Any compromises or hard choices that were made for a very specific set of reasons are usually worth documenting and protecting in patents.

- What information is need to recreate your design and process at a new facility?
- What is subtle, nonobvious, or counterintuitive?

  ○ What information is only in the head of the person who performed the work?
  ○ Sometimes when joining a team to work on an existing design, features and choices seem strange and illogical. But they were there for a reason—why so and what was their history?

- What tools or processes have changed over time?

  ○ You may need to be able to go back to a setpoint that worked for you earlier.

- How many design iterations have occurred and are the current design elements documented?
- How did cost, reliability, time to market, availability of materials, or manufacturability influence design choices?

With all of this in mind, the minimum set of documentation of your technology and product that should be maintained over time would include:

- Milestone design review documents and presentations.
- System functional diagrams and schematics.
- Simulation models and results.

- Photomask data with layer definitions (for MEMS, as well as for ASIC, if used) and CAD files for all non-MEMS components and assembly diagrams.
- Process runsheet, including start material specifications and process tolerances.
- Inspection requirements and pass/fail criteria.
- Package and test specifications, including bill of materials.

If you are struggling with motivating your team to properly document these items (it is very easy to procrastinate!), we recommend creating specific points in the development work schedule that have formal written reporting requirements, such as team meetings, formal design reviews at milestones, and so on. Making it a group activity will help your team to create consistent documentation for each of the above items[1].

The preceding bullet list itemizes the minimum set of documentation for your MEMS product. In the rest of this chapter, we will focus on the documentation required for a successful transfer of a MEMS design and process flow to a MEMS production foundry. If your product happens to have multiple MEMS components, you would need a set of documentation for each. High-level remarks on items that are peripherally useful to the foundry or necessary for fabricating non-MEMS components at other vendors are also provided.

Milestone Design Reviews

In Chap. 6, we discuss the value of milestone and go/no-go checkpoints in budgeting for MEMS development and in controlling costs along the way. Formal milestone review meetings provide great opportunities for documenting your team's knowledge because they require the team to create a presentation of the principle of operation and applications of your device, device specifications and requirements, and summaries of test results, and so on. The milestone review meeting becomes a forum for peer-review of the presented work.

The documents from milestone review meetings become a record which can later be mined for information needed to map the development of your final advanced prototype and which could also be provided to the MEMS foundry. The documents would also reflect the consensus within your development team regarding items such as critical design decisions, key processing issues and successes, and what constitutes a good device. You will appreciate your effort to create milestone summaries when, inevitably, later on, you will need to find a key piece of information from the development history.

---

[1] Consider implementing any of several commercial groupware or project management software options for facilitating team documentation.

System Functional Diagrams and Schematics

It is vital to create a top-level summary of how all system components interface with each other. At the start of a product development effort, this may be little more than a one-page document that outlines the proposed organization of your product and its components (Please see Chap. 9 for more information on components of a MEMS based product).

A system schematic should include all of the product's components: MEMS, PCB, ASIC, readout electronics, software, purchased components, etc., even if it is just a simple block diagram. The initial system schematic will help you to communicate to your entire team how all of the product components work together—what needs to be built or bought, which component interfaces with which and how, why component X needs to be able to do a, b, and c, etc. As the number of unknowns shrink during the development, and design specifics come into sharper focus, you must continually update these schematics. You may need to have different versions of the entire system schematic made for different audiences, especially if there is a need to compartmentalize information among vendors in order to protect trade secrets or confidential information.

Simulation Models and Results

Well-documented models serve an important purpose during transfer to manufacture. As your company's R&D process gets adjusted and translated to the equipment and methods available in the manufacturing setting, inevitably, the vendor will ask, "Could we change X to Y? It will make the process faster, cheaper, etc." You may need to revisit models in order to quickly assess the impact of any proposed changes.

Modeling efforts during development can spawn hundred, if not thousands, of computer files. It is too easy to create multiple design variations and, with time, you could quickly lose your ability to distinguish between model A and model D. Descriptive filenames can help, but they can also get cumbersome if too lengthy. Keeping a running list of the different model variations may help; however, an even better method is to create summary documents of each model, along with key input information that could be used to rerun the model, and a summary of prior model results.

You will likely have models for the complete product system, as well as individual components, in different software programs that are best suited for each type of device. In general, you want to document the model input and setup conditions, assumptions, results and conclusions. For illustration purposes, good documentation of a MEMS finite element (FE) simulation would include the following:

- Original input script file, with comments.
- Image of the model's physical geometry with any symmetry choices, modifications, etc. identified.

- Boundary conditions, element type, meshing choices.
- List of material properties data used.
- Input and output variables in table form.
- Images of complete and meshed 2D/3D models if applicable.
- Interpretation and summary of simulation results and list of conclusions that influenced design decisions (ideally, also captured in a milestone design review document).

This last item is helpful because the data may not always speak for itself a few months later, and for those less skilled at interpreting FE results, it can provide a clear understanding of why certain conclusions were drawn or design decisions made.

Documentation for MEMS Foundry Transfer

We now focus on the items needed to document the process of record (POR) for the MEMS device. You will need to provide a POR to your MEMS production foundry at the start of the foundry feasibility stage. The photomask layout data and the process flow description are just the basic documentation of your MEMS design. A well-organized and documented POR, however, uses consistent and clear terminology and includes a number of supporting documents. In this section, we outline some best practices and recommended supporting documents. If you have more than one MEMS device or are transferring your technology to more than one foundry, you will need to develop this document portfolio for each component and each foundry[2].

Photomask Layout Data

Photomask Layout File

The photomask layout data file is one of the two most important documents you provide to the foundry. It contains the design data that the foundry will use, along with the process flow, to manufacture your MEMS device. This data file may contain:

- A wafer-level layout, such as for a contact photomask.
- A group of identical die-level layouts, such as for a stepper reticle.

---

[2]The information transfer process will vary from foundry to foundry. The documentation guidelines in this chapter provide a strong foundation which you can later customize as needed.

- A single device design ("unit cell").
- A series of design variants.
- A series of test structures.

Clear communication with the foundry regarding your data file's contents is, therefore, critical. Your initial layout data file is the starting point of an important conversation between your company and the foundry. If your photomask data is not well organized, or not formatted as the foundry expects, confusion may lead to mistakes that will cost time and money.

Even if your company and the foundry happen to use the same layout software, you will most likely be transferring your data in GDSII format, not the software's native format. GDSII is the format used by photomask vendors to fabricate photomask reticles and all MEMS- and CMOS-specific layout editors can import and export it[3].

The following guidelines can help your foundry team to quickly and correctly understand your device layout[4]:

- Provide the foundry with copies of your photomask layout in both the layout editor native format (assuming the foundry is able to read it) and GDSII format.
- Always check your GDSII layout file for conversion errors before submitting it to the foundry (see Chap. 11).
- Add the date and/or version numbers to the photomask data filename, since revisions may be needed during the transfer process.
- If providing die-level data to the foundry, create a list of the top-level cell for each design variant (if any) or save each variant in separate files. This may also be needed for any test die.
- Include a picture of an example array of die to unambiguously illustrate the required die spacing, which determines the size of the die after wafer singulation.
- If providing a wafer-level layout to the foundry, provide a list explaining the hierarchy of all key cells[5] and a visual wafer map.
- Clearly identify data layers that have been biased to account for differences between the layout and the expected final device dimension. In general, we recommend providing unbiased data so that the foundry can make their own data bias adjustments according to their toolset performance.

---

[3] Other formats may be used by photomask vendors, but GDSII is the standard for MEMS. CAD programs that are not designed specifically for MEMS or CMOS may require a third-party program to convert data to GDSII format. Photomask vendors' websites can provide additional information, for example, see reference [1].

[4] These guidelines are also useful for communication between your design and fabrication teams during prototyping.

[5] Key cells are, for example, cells containing design variants or test structures. Do not include lower sub-cells such as one that contains only a single wirebond pad that is used to create an array of pads within a higher-level cell. More information regarding cell hierarchy can be found in Chap. 11.

Photomask Layer Definition

Once the foundry engineers read the photomask layout file, they will see a series of data layers for each corresponding lithography step in the process. The following guidelines can help to clearly communicate information regarding these layers to the foundry (as well as documenting for internal use during the proof of concept and advanced prototyping stages).

- Label all mask layers within your layout editor using clear, terse descriptions of the specific process step. For example, use Metal_1_Al and Metal_2_CrAu vs. just Metal_1 and Metal_2.
- When exporting mask data to GDSII, data layer names can sometimes be lost; therefore, provide a separate list of mask layer names and their corresponding GDS layer numbers (required for all GDSII files).
- Label the mask layers and corresponding GDS layer numbers in order of the process sequence.
- If processing on both sides of the wafer, establish a labeling and numbering system that clearly identifies to which side of the wafer the mask will be applied and from which vantage point the data is drawn (top or bottom view). (For example, use even GDS numbers for front side masks, odd GDS numbers for back side masks.) This helps to prevent mistakes in mask data parity[6] when ordering the photomask reticle, as well as during wafer processing.
- Keep layout data color schemes consistent across all photomask files and supporting documentation (e.g., use consistent colors in both cartoon device cross-section diagrams and also in mask layout data file). This helps to keep the foundry team from accidentally confusing features or layers.

The purpose of these guidelines is to not only help with the data transfer, but also to keep track of the data during the advanced prototypes stage, when the designs and process flows are in flux. Once the photomask layout data have been transferred, the foundry may have their own protocol regarding naming each layer and assigning associated GDS numbers. This is fine because they will have their own internal requirements for managing the mask data.

---

[6]When you are looking at a photomask layout in the layout editor software, you likely have a top-down view of all of the layers. However, some of these layers may be transferred to the back side of a wafer during processing. In that case, because the wafer will be flipped over, the mask data must also be flipped or mirrored. When ordering a photomask reticle, the parity of the mask must be specified to determine how the mask vendor will transfer data to the mask. Each mask vendor has their own format to convey mask parity information. See Chap. 11 for more information. If you have created a bottom-up view of data for a pattern that will be transferred to the back side of the wafer (not recommended), this must be clearly communicated to the foundry.

High-Level Process Flow Summary

We recommend that you also provide a high-level process flow summary accompanied by pictures of the layout, to act as a quick reference guide for the foundry's engineers. Ideally, this document would include both a top-down view of the photomask layout data and the corresponding cross-section diagram of the MEMS device after each process module has been completed.

Cross-section diagrams are a universal communication tool among MEMS engineers[7]. You likely have already created them for internal communications and milestone design reviews. They are also quite helpful when transferring photomask data to a foundry because they can unambiguously convey how the layout pattern data should be transferred to the wafer during processing. (Specifically, will an area drawn keep or remove material on the wafer.) See Fig. 16.1 for an example of how mask data can be drawn in two different ways and still produce the same feature on a wafer, and how providing a cross-section picture would clearly communicate the designer's intent.

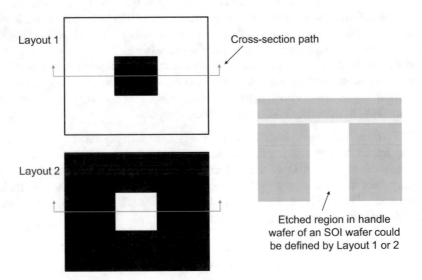

Fig. 16.1 Two different ways to draw photomask data that could produce the same final feature on a wafer. In Layout 1, the center square data is drawn (in black) to define the etch region, but would be digitized clear on the photomask for positive photoresist. In Layout 2, the data drawn (in black) show where silicon will remain post etch, and would be digitized dark on the photomask for positive photoresist

---

[7]You will find cross section diagrams used throughout MEMS design and process textbooks, such as refences [2–5], to illustrate devices and their process flows.

Layout Design Specifications

Layout design specifications convey aspects of the design that are critical to successfully fabricating the MEMS device so that it will meet your specifications. This information can come from models, as well as your prototype process and test experiences, and includes:

- Maximum allowable photomask manufacturing data grid size.
- Minimum feature size on each layer, also referred to as the "critical dimension"[8].
- Dimensional tolerances for device features, specifying which are crucial to the device performance.[9] Tolerances may be in the form of minimum and maximum values, or a plus/minus percentage of a target value.
- Layer to layer alignment tolerances essential for the device performance—often linked to specific features. This may be distinct from the alignment tolerance that you would specify for a given lithography layer.
- Layers that required data bias during advanced prototype processing and the magnitude of the bias that was applied.
- Any wafer- or die-level exclusions zones that are required for the MEMS process flow or subsequent back-end processes.

If you are using a foundry platform process for your MEMS design, all of these layout specifications will be provided to you in the form of a design rules document also known as a process design kit (PDK), and perhaps even a design rule check (DRC) dataset to interface with your mask layout software[10]. Your mask layout must conform to these foundry design rules, which have been set by the foundry in order to enable successful fabrication on their platform process. If you might need to violate any design rules, you need to discuss it with the foundry ahead of time; otherwise, expect the foundry to reject any nonconforming mask data.

MEMS Process Flow

Runsheet and Process Tolerances

The process flow runsheet (also sometimes called a runcard or traveler) is the other most important document you will provide to the foundry. Gather only the most current and relevant process information from the latest successful advanced

---

[8] The smallest feature size will determine the type of lithography and photomask reticle to be used.

[9] This data is the final desired dimension after processing is complete and is vital to accurately transferring a process flow. The foundry will need to determine what is required to achieve these tolerances.

[10] Earlier, we specifically did not refer to the layout design specifications as "design rules" because they have not yet been qualified for production purposes. The definition of a design rule is, that if you follow it, everything will work as expected.

prototype runsheet (see Chap. 11 for more information on creating a runsheet). Providing information on processes that you had previously tried and had also worked can sometimes be useful to the foundry as potential process variations that might be acceptable. You might also consider providing sample die and SEM or microscope images of each process step from prior prototype runs, in particular, for any features that are required for your product to work well. When the foundry inspects the process wafers, they will be able to reference these images.

Early in development, tolerance data and pass/fail criteria may be not as well understood, but as modeling, process, and test data are documented and compiled, a clearer picture of the process window that creates usable devices will emerge. By the time you are ready to transfer a process to a foundry, you have probably determined rough process windows for some but not all of your process steps. Table 16.1 provides an example format for a runsheet that is suitable for transfer to a foundry (also presented in Chap. 11).

During prototyping, the specific tool and recipe inputs were the most critical information for your fabrication team; now, it is the process observations gathered during prototyping which are most valuable for the foundry transfer. The foundry needs to know the specific values for alignment tolerances, geometry and inspection data, as well as any other details that make the process successful. Hopefully, all of those had been documented in the "Tolerances" and "Processing Notes" columns of your prototyping runsheet.

Importantly, the foundry is unlikely to have the same set of tools as your development fab did. Process notes from development provide important insight on what defines a successful process step, so that the foundry engineers can be sure to make their own tool deliver that same end result. As mentioned in Chap. 11, development runsheets can sometimes end up a disorganized mess of both extremely valuable and utterly useless data. We recommend that you don't leave organizing or mining this data until the last minute, when you are preparing for the foundry transfer. Developing the habit of doing a detailed runsheet review and cleanup after each process run will drastically reduce the time and effort required to get your compiled runsheet ready for the foundry.

Table 16.1 Example runsheet format to document the Process of Record (POR) which created the advanced prototype to be transferred to the foundry. (Note: rows are intentionally left blank)

Step #	Description	Tool	Recipe and Input Variables	Tolerances Pass/Fail Criteria	Inspection Details and Processing Notes	Wafer IDs
1						
2						

Inspection and Acceptance Criteria

The process tolerance and notes on the development runsheet, along with the mask design feature size and alignment tolerances, inform what needs to be inspected throughout the MEMS wafer process flow. It defines what will eventually become wafer acceptance criteria. If there are any specific surface cleanliness or environmental storage conditions that need to be maintained during production, these should also be listed as inspection criteria. Having consistent process and inspection documentation throughout development will speed up convergence on identifying the appropriate foundry acceptance criteria.

For acceptance criteria, it is critical to know which parameters define a successful product and what to measure to determine if your product is indeed "good." For example, would a mirror surface need to be *completely* particle free or would particles need to be below a certain diameter and count number per area? Identify any successful parts from the advanced prototype stage and study why and how they were successful. You must already know which process conditions and results you need in order to achieve sellable product before you could even ask the foundry to deliver them.

The list of wafer acceptance criteria is usually formed just prior to the start of the foundry pilot production stage, so that the foundry can focus on determining the process window to meet them. When compiling an acceptance criteria list, it is important to remember that the foundry will only agree to acceptance criteria which they can measure and control. For example, they will agree to holding a tolerance on the final die size, if the singulation process is a step they are managing. If that step is being managed by a back-end vendor (see next section), the foundry will only agree to hold a tolerance on the spacing of the die on the wafer, which is controlled by the foundry's lithography process.

Documentation for Other Supply Chain Vendors

Back End, Packaging, and Other Supply Chain Vendors

We have outlined the documentation needed for the foundry transfer in detail; however, this general approach can also be applied to documentation for any back-end[11] or other vendors. All system components that will integrate with the MEMS device to achieve the desired product performance should already have been documented during the prototype stages. Unless during prototyping you were already working with a production vendor, you will also need to transfer your back-end processes from development vendors (or your own internal team) to manufacturing vendors.

---

[11] "Back end" refers to any wafer-level steps executed after foundry fabrication is complete, such as wafer thinning, polishing or sawing.

For illustration purposes, we have created a list of possible back-end-of-line (BEOL) process documentation for a MEMS product. This documentation builds upon the schematics of the MEMS product components we discussed earlier in the chapter. We have not included every potential packaging variation in this list (please see Chaps. 12 and 13 for more background). Your documentation will vary depending on your specific requirements, number of components, packaging method, and materials. The more components and vendors you are using, the more critical the documentation becomes, as each supplier may handle the transfer differently. Possible BEOL processing documentation includes, but is not limited to:

- Bill of materials for all components and materials used in your MEMS product.
  - Vendor or manufacturer information for each component.
  - Component data sheets or schematics.
    - Component mechanical drawings, electrical schematic, including pin out.
    - Part numbers for off-the-shelf components.
  - Certificate of compliance and incoming inspection criteria for all purchased components or materials.
- Back-end inspection, test and assembly flow/traveler with supporting documentation:
  - Wafer probe method for MEMS/ASIC components.
    - Probe card schematic, layout, and vendor information.
    - Wafer map for probe tests.
    - Lists of inspection or function tests (see Chap. 20) and pass/fail criteria for each component and assembly step.
    - Optical inspection/tests (if applicable).
  - Wafer thinning.
    - Method, tech specifications, and inspection criteria.
  - Wafer singulation.
    - Method, tech specifications, and inspection criteria.
  - Package assembly.
    - Package mechanical drawings, electrical schematic, including pin out.
    - Part numbers for off-the-shelf packages.
    - Technical specifications for the following operations:
      Die attach adhesive dispense.
      Die attach process.
      Wire bond, bumping, solder, etc.
      Final packaging (e.g., lidding, glob top, gas backfill, over-molding).
    - Inspection methods (e.g., electrical or optical) and pass/fail criteria.

- ○ Package-level function test and inspection.

  - – List of tests and method for each test.
  - – Pass/fail criteria.
  - – Electrical calibration and compensation methods.

- ○ Shipping requirements.

  - – Labeling.
  - – Format for delivery to next vendor or your company (e.g., containers, trays or reels, wrapping materials).
  - – Shipping documentation, customs forms, etc.

It is important for your company to maintain internal copies for as much of this documentation as possible. Like production foundries, most vendors are unlikely to provide information on exact process details, tool settings, etc. as this information can be considered proprietary. You should still know the basic parameters for what each vendor is doing, in case you might later need to find a second source vendor.

Wherever possible, acquire copies of all operation travelers (the document that provides information on what and how an operation is to be executed, also sometimes called a runsheet or runcard[12]). If a vendor is unwilling to provide the traveler, record the specification and instructions which you provided to the vendor, and at the very minimum, get the filename and version number of the traveler used for the operation. This will allow you to reference it on any future jobs and to confirm the correct operation was selected for execution.

Process Quality Control and Qualification Test Documents

The process and operations data described in the above sections should be compiled within a formal inspection or acceptance criteria document or portfolio. This is an important step in to begin building up quality control documents within your company. If you might be developing an FDA-controlled medical device, be aware that any development and manufacturing documents will have to adhere to the FDA's documentation requirements.

Your process quality control document can be as simple as a check list, such as our example in Table 16.2, or it may be expanded to include documentation of every step of your product build operations, including reference data (e.g., typical or historical data) and step-by-step photos.

Table 16.2 Example check list for inspection/acceptance criteria (note: rows left intentionally blank)

Description	Vendor	Criteria (typ, max, min)	Reference Document
Step 1			
Step 2			

---

[12] In our experience, BEOL vendors usually use the term "traveler."

To avoid confusion, the inspection and measurement data outlined in this chapter are manufacturing process quality data only, not product qualification test data. This latter topic is described in more detail in Chaps. 14 and 20.

Briefly, qualification testing is done to the final product to ensure that it can meet all of the product requirements as well as any regulatory or established standards. Qualification tests are highly dependent on the product type (e.g., accelerometer, pressure sensor, scanning mirror), the type of end use (e.g., medical, automotive, consumer) and the associated reliability requirements, and the operational and storage environment (e.g., temperature range, harsh or dirty environment).

During development, the product qualification tests may be repeated, as design iterations occur, in order to achieve the specified product requirements. The following information from qualification test should be documented using version control.

- Product requirements, by customer or end application.
- Lists of required standards.
- Lists of tests (environmental, shock, lifetime, etc.)
- Method for each test.
- Pass/fail criteria and test results.
- Certification of qualification.
- Certifications of all vendors/facilities in supply chain.
- History of customer returns/complaints.

Summary

Documenting your MEMS technology and product specifications during development is a critical but sometimes neglected task. Documentation is a prerequisite to successfully transfer your product from a development environment to manufacturing one. Establishing documentation protocols early in development, as well as understanding what will be required for the transfer to manufacturing, can help keep this important task high on your company's priority list and keep it from becoming burdensome. When you are finally ready to transfer your technology, the documentation you created for each component during development will be a starting point for a relationship with each foundry or supply chain vendor. Chap. 19 provides more information on the actual transfer process to begin production at a MEMS foundry.

References

1. Compugraphics (2017) Education center. https://www.compugraphics-photomasks.com/education-centre. Accessed 5 Aug 2020
2. Kovacs GT (1998) Micromachined transducers sourcebook. McGraw-Hill, New York

3. Madou M (2011) Fundamentals of microfabrication and nanotechnology, 3rd edn. CRC Press, Boca Raton
4. Maluf N, Williams K (2004) An introduction to microelectromechanical systems engineering, 2nd edn. Artech House, Boston
5. Senturia SD (2004) Microsystem design, 2nd edn. Springer, New York

Part IV
Technology Transfer and Scaling Up Manufacturing

Chapter 17
Determining Readiness for Volume Production

As development of a MEMS product progresses, executives and investors may begin to apply well-intentioned pressure to move the fabrication process from a R&D environment into a production facility (fab or foundry). Moving processes to a commercial manufacturing environment, where the MEMS product will be made in high volume, is an essential transition for product commercialization. For anyone funding product development and eager to start selling product in order to recoup an investment, this transition can never come soon enough.

When is the optimal time to make this crucial transition? Moving technology prematurely to the production environment could actually slow down development progress, if the development team still needs small R&D wafer orders which could then get de-prioritized in the bustle of a production fab. Waiting too long to move can also be problematic, if maturing a MEMS technology might be handicapped by inferior performance of a R&D fab's aged fabrication equipment. Determining readiness to move to production recalls the children's tale of *Goldilocks and the Three Bears*; the porridge should be not too cold, not too hot, but just right. In MEMS, determining when is "just right" depends on both technology and business factors unique to the product and company. In this chapter, we will illuminate those factors and discuss how to evaluate readiness to move.

Understanding Development Versus Production Environments

It is worthwhile to first gain a deeper understanding of the similarities and differences of MEMS development fabs versus production fabs.

Development fabs are smaller facilities, usually less than 1000 sq. m. in area, and tend to use smaller diameter wafers for fabrication than production fabs. As of this writing, most MEMS development fabs operate using 150 mm wafers, with a few having capabilities for 100 mm and/or 200 mm wafers. Most MEMS volume

Table 17.1 Diameter of wafers in use today by MEMS fabs compared to years in common use by the semiconductor industry [1]

Wafer diameter (mm)	Years in common use by semiconductor industry
100	1975–1990
150	1980–1990
200	1990–
300	2001–

production fabs operate using 200 mm wafers, with a few smaller production fabs using 150 mm wafers. As of 2020, high-volume MEMS production fabs operate exclusively on 200 mm wafers; a few are just starting to move up to 300 mm wafers.

Because most equipment companies originally engineered their process tools to serve the semiconductor industry, which systematically increased wafer diameters over the years, one can roughly estimate the vintage of the equipment at a MEMS fab according to wafer diameter.[1] Therefore, when it comes to process technology, MEMS development fabs can be one generation behind the MEMS production fabs, which then in turn can be one generation behind the semiconductor fabs. See Table 17.1.

In general, the main challenges with older process tools, even if they are well maintained, is across-wafer process uniformity and wafer-to-wafer process repeatability. Older tools, however, tend to have options for running in manual or semiautomated modes versus state-of-the-art, automated high-volume manufacturing tools. Older tool sets will therefore provide more process flexibility, which is very valuable during R&D, in exchange for wafer uniformity and yield, which is less critical anyway in early stage development.

Production fabs are organized around maximizing manufacturing cost efficiency and wafer throughput. This focus is especially sharp in CMOS semiconductor foundries which on occasion may do volume production of some types of MEMS. High-volume fabs are at least 10,000 sq. m in area, filled with automated tools, and minimize the use of human operators. At production fabs, the newer tools will offer much better process speed, uniformity, and repeatability than earlier generation tools, because of more advanced process technology. Most 200 mm tools can only be run in semiauto or automatic mode and require specialized, sealed wafer cassettes known as Front Opening Unified Pods (FOUPs).

Some MEMS production fabs have partnerships with development fabs that act as "feeders" to the production fab. The development fab partner may be a nearby government-funded research facility. The partnership aims to help streamline the eventual transfer of a new product technology to the higher volume production fab.

---

[1]A few equipment makers have kept selling tools for particular wafer diameters long after the semiconductor industry moved up to larger wafers, so for some types of process equipment, it is still possible to purchase new 100 mm, 150 mm or 200 mm tools.

Staff Composition and Mindset

The differences in staff composition at development fabs versus production fabs reflect the different purposes of these respective facilities. A development fab, being focused on discovery and development, tends to employ more process engineers, and relatively fewer operators or technicians. The process engineers usually have multidisciplinary expertise in all of the fab's processes, understand MEMS process integration, and sometimes execute wafer processes themselves. The process engineers will also interface directly with the customer's engineering team. The operators at development fabs also tend to be quite versatile and have multidisciplinary skills to operate many tools.

In a production fab, the engineer-to-operator ratio is inverted, and all technical staff are narrowly specialized. The majority of a fab's staff are tool operators, focused on efficiently executing the consistent tasks of volume manufacturing. Tool operators are specialized by process module, such as lithography, diffusion or etch, and tend to not have interchangeable duties. Because the facility's focus is production, there are proportionately fewer process engineers, and of those, most are engineers specializing deeply in just a single process area, such as lithography. A production MEMS fab will have only a handful multidisciplinary engineers who are knowledgeable about MEMS design and process integration. A CMOS fab which also accepts some MEMS devices may not have any experienced MEMS process integration engineers on staff.

Typically, the engineers in a production fab do not process wafers themselves nor do they interact with the customer, other than at the start of a project or if a problem might arise during production. The customer's main point of contact at a production fab is a project manager, who may or may not be a MEMS engineer.

Another significant difference between the two environments is mindset of the staff. Each of the two fab environments select for a different type of optimal staff mindset. The staff at a development fab are likely to be creative, willing to experiment and have good tolerance of failure, and be less concerned about detail or process repeatability. In contrast, the production fab personnel are likely to be detail-oriented, methodical and inclined to safeguard status quo, and reluctant to change quickly or to take creative leaps. Each respective environment selects for and reinforces its preferred employee mindset and values it as essential for business success. When transitioning your company's MEMS device from a development to a production fab, these vastly different mindsets might give you some initial culture shock. [2]

Job queuing and Process Stability

Job queuing is the art and science of squeezing the most value from a facility's fixed tool set and capacity. A fab is composed of a diverse set of tools, some of which may have only single wafer process throughput, such as DRIE; some that allow simultaneous processing in batches of 6–9 wafers, such as PVD; or some that allow batch processing in groups of 100–200 wafers, such as furnace oxidation. The fab manager's job is to figure out which customers' wafers to queue at a tool, in which order, so that all tools are kept at high utilization and wafers do not start to pile up while waiting to be processed. It can take millions of dollars and several months to purchase and install new tools in a fab and then qualify them for production, so tool and overall fab capacity are carefully tracked and managed [3].

In a development fab, job queuing may be done manually or with simple software tools, like a calendar program or an Excel spreadsheet. Since development fabs tend to have older, depreciated or less expensive tools, there will be less financial pressure to keep tool utilization extremely high in order to maintain overall company profitability. Depending on the customer mix at a given time, some tools in a development fab may comfortably remain underutilized.

Due to reduced pressure to maintain high tool utilization, development fabs therefore have much more flexibility in job queuing in order to attempt experimental processes. Changing a tool's process settings or introducing new recipes makes a tool temporarily unavailable to other jobs, and furthermore creates a risk that returning the tool to its original performance might not go smoothly. Before a tool can be returned to normal use, its settings need to be reset and then tested and validated; some wafer processes are multivariable and fragile, requiring iterative tuning. This can add up to significant tool downtime (days or even weeks), which further interrupts job schedules. Development fabs can more easily tolerate these disruptions and so can be more accommodating of a customer's request for process experiments.

Production facilities, on the other hand, are narrowly focused on maintaining high tool utilization and wafer throughput and are understandably averse to any changes which might cause job queuing disruptions. Production fabs are filled with much more expensive, newer tools that carry financing costs. Tool utilization must therefore be kept as high as possible in order to maximize profit, which is needed not only for funding company operations but also to pay off the capital expense of the tool set. In order to achieve high tool utilization, job queuing must be highly optimized and tool risks must be minimized. Production fabs use specialized software, known as MES or Manufacturing Execution Software, in order optimize job queuing.

Tool operations are organized to minimize tool downtime, which means that only a limited number of documented, highly tested and stable processes are allowed to run on a tool. Operators and engineers continuously monitor the performance of the processes and record data on stability. Any contemplated change to a tool or its recipes would therefore need to be carefully evaluated, receive formal authorization, and then be scheduled. Impromptu process changes are simply not possible in this environment. Any last minute requests by a customer, such as for mid-process changes or wafer batch splits for A-B option testing, will cause the customer's wafers to be pulled out of queue and set aside until one of the foundry's few engineers can evaluate the situation. In other words, in a production fab, any deviation from a planned work order will certainly result in delays.

Business Model

A development fab, being focused on R&D, may earn most of its revenue on project-based fees or nonrecurring engineering (NRE) fees. These fees may be charged as labor time and materials (T&M) consumed during the project or as a fixed, milestone-based fee. Development fabs may also have a mix of other business types, including one-step process services and high-value, low-volume production work. MEMS components for medical devices or defense applications are often made at smaller development fabs due to the low annual wafer volumes of these specialty products.

The production fab's business model is focused on earning revenue by selling wafers, so its goal is to get the customer to volume wafer production as soon as possible. Foundries will charge new customers NRE fees for the feasibility or pilot batches of a new product, in order to cover costs while they learn the new process and make adjustments to improve its yield.

A foundry may consider discounting the NRE charge in order to entice a customer, if they assess that the customer is highly likely to eventually order large numbers of wafers. In that case, the NRE costs would be amortized over the wafer price and later recouped in volume wafer sales. Only customers that are well-established companies and already have other products in the marketplace, strong evidence of their competency at commercialization, would enjoy this benefit. Start-up or less-established companies are most likely to be charged full NRE, simply because savvy foundries know that the odds are low for a young company to successfully surmount all of the challenges of growing itself while also developing a market for a new high-volume product.

The foundry business model is most profitable when a stable process has been running for years, tool utilization is high, and the fab runs 24/7. As the years elapse, the foundry continually works on improvements to streamline a process. Customers whose products bring the foundry closest to this optimum operating point will receive the most favorable pricing.

There is a common misconception among new entrepreneurs that "everything will get cheaper when we move to a production fab." The discounts associated with economy of scale, however, cannot be achieved until high-volume manufacturing actually occurs! For the reasons described above, the first few years of MEMS fabrication at a production fab are likely to cost more than they did at the development fab.

Evaluating Readiness for Transfer

Assessing readiness to transfer to a production fab comes down to understanding the pros and cons of the development versus production fab environments, doing a clear-eyed assessment of where your product development stands, and then determining the best environment for achieving your near-term goals [4]. To be ready to move to a production fab, the following minimum technical and business requirements should be met.

Technical Requirements

To be technically ready for transfer, all research on how to fabricate an adequately functioning device should be complete. There should be no need for more process experiments to understand how the device works or how to optimize its performance. The relationships between process parameters and process tolerances and device performance must be well understood. The baseline process must be fully documented and stable, although it does not have to be high yielding. In other words, every process step needed to build the product must have defined performance tolerances (i.e., pass criteria) and also an understanding of which tolerance excursions will cause it to fail.

A critical need for higher performing tools and/or access to enabling foundry IP, such as TSVs or specific process modules, are compelling and valid reasons to move to a production fab prior to reaching the readiness level described above.

Other technical considerations for moving to a production fab sooner would include having a need to produce large numbers of sample devices for product reliability testing, or if the opportunity cost of remaining at development fab is too high, such as due to inadequate throughput of wafers caused by lack of tool automation.

Prior to beginning the transfer, your company should also have available one or more engineers who know the product's fabrication process well and who can support the production fab engineers, whether by regularly scheduled phone calls or extended visits. The production fab will need regular support during the first few wafer runs in order to learn the new process as quickly and cost effectively as possible.

Business Requirements

As of time of writing, when transferring a process to a 150 mm MEMS wafer foundry, expect to spend at least US $500,000 per year just for foundry services; additional budget will be needed to similarly transfer electronics, back-end processing, packaging, etc. to respective production facilities. For transfer to a 200 mm MEMS wafer foundry, expect to spend at least US $1,000,000 per year just for foundry services. These numbers are representative for a medium-complexity MEMS device (6–8 mask levels). Annual costs will vary according to the complexity of the device and the total number of wafers processed.

Having adequate funding is an absolute requirement for a company to be ready for transfer to a production fab. Do not begin a transfer without at least 2 years' worth of budget reserved. Long interruptions in wafer processing due to waiting for more funds will jeopardize the "freshness" of the process, result in the foundry's engineers being assigned to other customer projects, and ultimately waste any earlier efforts. Once the production fab starts learning a process, for best value and outcome for all parties, the fab should be able to build upon and improve their process knowledge without interruption.

Other Supply Chain Vendors

Although this chapter focuses on describing readiness factors for transferring the MEMS wafer fabrication to a production facility, the same guidance is valid for other components of the MEMS product, such as ASIC wafers, board electronics, testing, chip packaging, system integration, calibration, and reliability testing. As with MEMS wafer fabrication, there is a distinct shift in mindset, operating principles, automation, and cost between development versus volume production facilities.

For each manufacturing operation in your product, you must similarly conduct a frank technical and business assessment of the pros/cons of the development environment versus production environment for each respective vendor and then assess your technology's readiness to transition from one to the other.

Special Case: Low Annual Wafer Volume

To a volume production fab, a customer's order consisting of less than 600 wafers per year (50 wafers starts per month) would be unattractive; an order of less than 100 wafers per year would almost certainly be rejected. Some foundries are less sensitive to the number of wafers and instead focus on whether your company can pay "table stakes"—a minimum annual amount.

Annual product unit volumes of less than one million chips, especially if the chip size is small (<4 sq. mm), correspond to low-volume wafer production. (See Chap. 5 for information on how to estimate the number of wafers needed per year to support a given annual product unit volume.) MEMS products having low wafer volume are those used in specialty, high-value market applications, such as medical devices, industrial machinery, scientific instruments, and military or aerospace vehicles. It is not unusual for production of a MEMS pressure sensor, to be used in specialized cardiology instruments, to require less than twenty-five 150 mm wafers fabricated per year, or for an aerospace vehicle sensor to need less than five 150 mm wafers fabricated per year. These extremely low numbers of wafers are untenable for a production fab, for the reasons described above.

In these specific product cases, there will likely not be an opportunity to transition to a production fab. Instead, anticipating this situation, one should endeavor to work at a development fab that will be suitable for both the early stage R&D and later stage product wafer fabrication. Some development fabs do carry ISO-9001, ISO-13485, and other typical manufacturing quality certifications.

Realistically, a low annual wafer volume means that the ideal manufacturing state of having a fully characterized, stable, cost-efficient production may never be reached. A fab needs to run the same process continuously, with large numbers of wafers, in order to get statistically valid process control data, to keep wafer yields high, and to keep per wafer costs as low as possible. When wafer runs are infrequent, every new batch started is like doing a small development project all over again.

This does not mean that products with low annual wafer volumes are doomed to low manufacturing quality. It does mean, however, that the expectations of investors, executives, and even designers need to be managed to what is realistically achievable in low volume. Development fabs can certainly manufacture high-quality products; however, expectations must be tempered on achievable wafer yields and on per-wafer cost, relative to the expectations of a high-volume production fab. More testing and evaluation would certainly be required in order to identify the good die made in a process that has not been tightly controlled. The effects of lower yield, combined with a need for more testing to identify KGD, further drive up costs in low volume production.

In conclusion, a low-volume MEMS product business can still be quite profitable and successful, as long as the costs and risks of low-volume manufacturing at a development fab have been fully understood and properly accounted for.

Summary

Determining readiness to move from a development facility to a volume manufacturing facility needs thoughtful assessment. There is no one-size-fits-all answer due to the variety of MEMS product types, wafer volumes, process complexity, and market conditions. A diligent analysis requires an understanding of the differences and features of R&D versus production facilities and the benefits each might confer at that stage of product development. A frank assessment of your product's technical readiness, as well as of your company's business readiness, goals, and capitalization, is needed. Some low-volume products may never be able to move to a production facility due to low annual wafer volumes.

Next in Chap. 18, we will describe how to best select a production facility.

References

1. Hattori T (2015) From 20 mm to 450 mm: the progress in silicon wafer diameter nodes. Telescope Magazine, April 30, 2015 https://www.tel.com/museum/magazine/material/150430_report04_03/
2. Petersen K (2014) MEMS entrepreneurial perspectives. In: Solid-State Sensors, Actuators and Microsystems Workshop Hilton Head Island, SC, June 2014
3. Wood SC (2000) Factory Modeling. In: Nishi Y, Doering R (eds) Handbook of semiconductor manufacturing technology. Marcel Dekker, New York
4. Fitzgerald AM, Jackson KM, Chung CC, White CD (2018) Translational engineering: best practices in developing MEMS for volume manufacturing. Sens Mater 30(4):779–798

Chapter 18
Selecting a Foundry Partner

Bringing a new product technology to full volume production at a foundry is an undertaking which requires years of work and millions of dollars. Because of this huge investment, one of the most important decisions a fabless MEMS product company will make is choosing its foundry partner. Just as in one's personal life, investing time and effort to find a well-matched partner will save much heartache and money in the long run.

Assessing Compatibility

When considering foundries, your first thoughts might naturally focus on seeking those with compatible technical capabilities or prior experience with a specific MEMS technology. In our experience, however, business and cultural fits are also crucial considerations to achieve a successful working relationship. Preparing MEMS products for volume manufacturing can be a difficult and lengthy undertaking, fraught with unexpected setbacks and challenges. Business and management culture compatibility, and ultimately strong commitment to mutual success, will help the partnership to function well and succeed over the long term [1, 2].

Business Compatibility

To initially assess business compatibility with a foundry, check that your company's expected annual wafer volume is compatible with the foundry's business model. By annual wafer volume, we mean the number of finished wafers your company will realistically buy from the foundry within a calendar year. (Please review Chap. 5 for more information on how to estimate the annual wafer volume needed to yield a given product unit volume.)

© The Author(s), under exclusive license to Springer Nature Switzerland AG 2021
A. M. Fitzgerald et al., *MEMS Product Development*, Microsystems
and Nanosystems, https://doi.org/10.1007/978-3-030-61709-7_18

As of writing, most MEMS foundry colleagues would consider 600 wafers (50 wafer starts per month) or fewer per year to be "small" or "low-volume" production and more than 10,000 wafers per year to be "large" or "high-volume" production. Orders smaller than 600 wafers per year will be unattractive for most production foundries and may require fabrication at a smaller foundry or a development fab instead (see Chap. 17). Orders of this magnitude, however, may sometimes be accepted under special conditions. Some of those conditions are:

- If your MEMS product will utilize the foundry's platform process, IP or multi-project wafer (MPW) program;
- If you are willing to pay a premium price for the wafers;
- If your MEMS technology happens to be very strategic to the foundry's business.

Ideally, your company should be neither the smallest nor the largest customer of the foundry. If your wafer order is much smaller than average, in busy times, it could become neglected or de-prioritized, while the foundry gives preference to customers with larger orders that generate more revenue. Your order could possibly suffer months of delay.

If your wafer order is uncomfortably large for the foundry, then delays could occur due to insufficient wafer throughput, or quality could suffer because the foundry's staff is overwhelmed by the large order. Ideally, order size should be in the range of 5–15% of the foundry's total annual wafer capacity.

A second dimension to business compatibility is having good alignment on market strategy. Foundries make their own assessments on which end use markets will drive wafer volumes and then develop a strategy to offer the key process technologies needed for MEMS devices sold into those markets. For example, one past foundry strategy might have been to offer high aspect ratio, high-throughput DRIE, needed by MEMS inertial sensors selling into smartphones, while another foundry might have chosen to offer a thin single crystal silicon layer transfer process, useful to make pressure sensors or capacitive micromachined ultrasound transducers (CMUTs) for medical devices.

It is important to gain insight into a foundry's strategy for the present and upcoming years. If your company's target market aligns with, or is the same as, the foundry's target market(s), your product will be regarded as "strategic" to the foundry's business. Some of the benefits of this special status are that the foundry will put extra effort into developing key process recipes or perhaps purchase a specialized tool in order to make your product, because they see a larger, longer term opportunity to reuse the recipe or tool for their other customers in that same strategic market. Strategic customers also help foundries to win even more customers by being evidence of the foundry's track record with a particular technology.

Management Culture Compatibility

When we refer to management culture, for the avoidance of ambiguity, we do not mean ethnic or language culture. Here we define management culture as a set of preferences on items essential to project execution, such as: communication style;

attitude towards failures; formality of documentation; tolerance of technical or budget risk.

Incompatible management cultures can create challenges, especially in the face of dealing with tough technical problems. As an example of how different cultures can result in friction:

- If your company is very informal in communications, and the foundry is not, then in a challenging situation, your company's informal tone may be perceived by the foundry as sloppy or disrespectful and could cause further ill will;
- If your company is very data-driven and thinks collecting and analyzing even more data is a good use of resources, then a foundry team with a more practical, results-oriented approach could feel micro-managed by many requests for measurements, and furthermore could be tempted to withhold information in order to avoid additional requests they perceive as burdensome; or,
- If your company is risk-tolerant, but the foundry is not, then your request of "let's just try it and see what happens" could be viewed as reckless and wasteful by a cautious foundry team.

These examples of incompatible management culture will grate on a business relationship over time. A crisis would amplify those incompatibilities and could result in serious damage to the partnership.

Failure and setbacks are a normal part of any product development, especially in MEMS. Having compatible management cultures will help both parties to work together to solve problems instead of devolving into accusations and mistrust. During the foundry selection process, you will need to spend time to meet the foundry personnel, learn about their business and management culture, and frankly assess your mutual compatibility.

Technical Compatibility

As mentioned earlier, technical compatibility between the foundry's toolset and skills and the process needs of your MEMS product is an obvious requirement and the first one most people focus on when evaluating potential foundry partners. There are, however, additional important aspects of technical compatibility to be considered.

Experience with a comparable level of process complexity, in terms of the number of process steps and the process tolerances, is an important aspect of technical compatibility. For example, the fundamental process steps used to make a microfluidic chip are the same as used for making a MEMS gyroscope: thermal oxidation, lithography, DRIE, metallization, wafer bonding. So at first glance, one might be lulled into thinking that any foundry which has the process equipment to make microfluidics could also make a gyroscope. However, the technical skill level required to fabricate those respective MEMS devices is quite different.

A microfluidic chip may have only 2 or 3 mask levels, whereas a gyroscope could require 6–12 mask levels, depending on its architecture. The gyro process would therefore require at least 2–4 times as many steps to execute, measure, and inspect correctly. The cumulative probability of the occurrence of defects multiplies over the number of steps. This means that foundries that will be executing long, complex processes must have rigorous inspection and quality control techniques in order to make sure any defects would be caught and corrected quickly mid-process.

Continuing this example, the process tolerances required to yield a functional microfluidic chip are quite different from those that make a good gyroscope. The dimensions of a microfluidic device may only need to be precise to ±5.0%, whereas a resonating gyroscope, whose proper function relies on geometric precision, would completely fail unless dimension tolerances could be held to less than ±0.5%. Controlling a DRIE recipe so that it can etch to ±5.0% tolerance is straightforward, whereas achieving the dimensional precision needed for a gyroscope will require significant tuning of the DRIE recipe, and possibly weeks to months of process validation work by an expert engineer.

When considering technical compatibility, it important to be sure the foundry has appropriate skill level with the relevant processes and has experience fabricating either the same type of device as yours, or one of comparable (or higher) complexity. If the foundry's skill level is too low, you will be paying in time and money for them to improve their skills on your project, if they can succeed at all.

Another aspect of technical compatibility is being receptive to considering each other's ideas and willingness to collaborate to solve technical problems. A "not invented here" bad attitude, in which someone automatically judges any idea that did not originate from them as inferior, is dangerous in either party. MEMS process problems can be complex and often there are multiple valid pathways to overcome them. Forceful insistence that "unless we do it our way, it won't succeed" will just devolve into a time-consuming battle of egos versus productive problem-solving.

Compatibility Considerations Important to the Foundry

There are additional considerations that foundries may take into account when assessing fit with potential customers. Due to the complexities of operating a fab (described in Chap. 17), foundries prefer fabricating products that will offer the stability of a long production run. In other words, MEMS products that will be manufactured consistently for five or more years are strongly preferred over products that will have only a 1 or 2 year run. It is harder to amortize ramp-up costs over a short product run. For example, automotive and medical sensors, due to the long initial product qualification in those respective markets, are often designed for a 5–10 year product run. The prospect of predictable revenue from running the same process continually for 10 years is highly desirable to a foundry. In contrast, a MEMS component going into a consumer electronics product might have a short run of less than a year due to evolving consumer tastes. The huge effort for a foundry to spool up for a one-time, short-run production is only tolerable if there will also be a large volume of wafers or a high profit margin in that production.

Furthermore, MEMS products that have the ability to maintain higher average selling prices (ASPs) over time are also strongly preferred over products subject to constant price erosion. MEMS used in high-value products, such as a surgical device with a price of US $1000, may have a stable ASP of $30 per die for years. At the other extreme, the MEMS components used in smartphones typically sell for less than US $0.10 each. With a billion smartphones produced per year in a very competitive marketplace, cutting the ASP of a MEMS die by a penny would save an OEM US $10 million per year. As a consequence, a foundry making MEMS components for smartphones can expect to be uncomfortably pressured, on an annual basis, to reduce its fabrication costs. A foundry that is in a strong financial position has latitude to choose its new customers. Such a foundry will not be so interested in making a commoditized MEMS component that would be subject to price erosion, such as a low-cost microphone, unless the opportunity might be balanced by some other highly appealing factors.

The AMFitzgerald Method for Selecting a Foundry

In our business, we have created our own methodology for conducting thorough foundry diligence in order to help our clients to choose an optimal foundry partner [3]. Figure 18.1 outlines the major steps and the flow of our method. Each block in the foundry selection process is described in detail in the sections below.

Timeline for Foundry Selection, Transfer, and Ramp-up to Production Readiness

Many clients have asked us, "When should we start a foundry selection process?"

First and foremost, your product technology must be mature enough so that a transfer to foundry is both sensible and achievable. (Please review the criteria for readiness described in Chap. 17.)

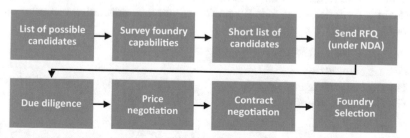

Fig. 18.1 Flowchart of the major steps in AMFitzgerald's foundry selection process. The steps in the first row identify the most qualified candidates and the steps in the second row evaluate the candidates, and finally enable selection. (Source: AMFitzgerald)

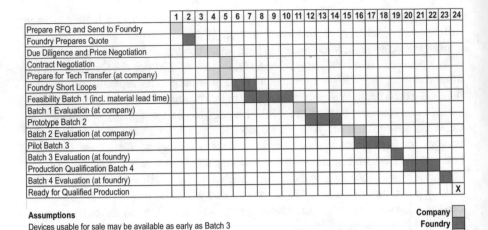

	1	2	3	4	5	6	7	8	9	10	11	12	13	14	15	16	17	18	19	20	21	22	23	24
Prepare RFQ and Send to Foundry																								
Foundry Prepares Quote																								
Due Diligence and Price Negotiation																								
Contract Negotiation																								
Prepare for Tech Transfer (at company)																								
Foundry Short Loops																								
Feasibility Batch 1 (incl. material lead time)																								
Batch 1 Evaluation (at company)																								
Prototype Batch 2																								
Batch 2 Evaluation (at company)																								
Pilot Batch 3																								
Batch 3 Evaluation (at foundry)																								
Production Qualification Batch 4																								
Batch 4 Evaluation (at foundry)																								
Ready for Qualified Production																								X

Assumptions

Devices usable for sale may be available as early as Batch 3

This is the minimum number of batches required to reach qualified production; in practice, many more are needed

Company   Foundry   Milestone X

Fig. 18.2 Timeline of the major events involved in selecting a production foundry, then transferring MEMS fabrication from an R&D facility to the foundry, and then ramping up to production readiness. (Source: AMFitzgerald)

Figure 18.2 outlines a typical timeline for the complete journey to production readiness: foundry selection, technology transfer, and manufacturing ramp-up. On average, we expect a thorough foundry selection process to take up to 5 months. The following sections of this chapter will describe what happens during those 5 months in more detail.

The technology transfer to the foundry and manufacturing ramp-up generally will take an additional 18 months, for a total timeline of 2 years. (Please refer to Chap. 19 for more information on how to execute a technology transfer and to Chap. 3 for a description of the typical Advance Product Quality Planning gates which most foundries use for manufacturing ramp-up.)

Two years may seem like an eternity, especially to a start-up company; however, one needs to consider not only the time needed to select the foundry, but to actually transfer the technology to the foundry. After transfer, the foundry has a lot of work to do in order stabilize the fabrication process and then build up the wafer volume in order to achieve full volume production capability.

Could it be done in less than 2 years? It is possible, depending critically on the following factors:

- How mature your MEMS design is.
- Process complexity.
- The skill of your team.
- The skill of the foundry.

If you have a simple MEMS device based on a well-established or already-qualified process (such as a foundry platform technology), if your team is full of seasoned engineers, and the foundry has made this same type of product before, it is possible to execute a technology transfer and reach volume production within a year. Without all of those factors in place, it is safest to plan for 2 years.

Build a List of Candidate Foundries

At the time of writing, more than 60 MEMS foundries exist in the world, of widely varying sizes and capabilities. It would take an unreasonable amount of time to evaluate each and every one, so your first temptation might be to call some trustworthy MEMS colleagues and ask for their recommendations on which foundry to pick. This is a good way to develop a list of candidates to consider, but not to actually select the foundry, due to all the pair-wise compatibility issues described above. A foundry that worked well for your colleague's company may not be the best fit for your company's needs.

Our method begins by creating an initial list of possible candidates, and then using two steps, systematically filters that list into a short list of highly qualified candidates, worthy of in-depth evaluation. The first step consists of a nonconfidential survey of high-level capabilities and interest, and then the second step provides a detailed and confidential Request for Quote (RFQ), sent only to the most promising candidates under NDA. This two-step process furthermore helps to protect your company's confidential information by limiting its exposure to only the most qualified candidates.

Survey Foundry Capabilities

We begin with a nonconfidential survey in order to gather current information on the capabilities and business interests of the initial list of foundries. Although you may already have some familiarity with one or more foundries on your list, or perhaps even had prior direct work experience, nothing is static in the foundry business. Capabilities change, people change, and business models change with time. Your or your colleague's knowledge of a foundry from an experience from 2 years ago may already be obsolete.

We recommend making the initial survey nonconfidential in order to remove the time, cost, and friction of signing a nondisclosure agreement (NDA). The survey is a simple tool that enables a high-level assessment on whether a foundry merits a closer look.

The survey should also provide the foundry with some basic information about your company's MEMS product, in order to assess three key factors:

- **Process compatibility:** Does the foundry already have the processes needed to fabricate your MEMS design? Are the processes in-house or outsourced?
- **Wafer size and annual volume:** Can the foundry handle the wafer size and volume your product will need?
- **Interest in the opportunity:** Based on some nonconfidential information you share about your MEMS design and its production needs, how appealing is this opportunity to the foundry?

The survey should be a short document that the foundry sales team can respond to quickly. Send the survey to as many foundries as appear to meet your needs and are in regions or countries where your company is willing to do business. The initial list of candidates might be composed of 10–15 foundries.

As the surveys are returned to you, it is a good idea to have follow-up calls with the foundry's salespeople in order to clarify survey responses and gather more information. Spend some time talking to the most promising foundry candidates to verify their interest and to tell them a little more about your manufacturing plans.

After analyzing the survey responses, you will find a cluster of candidates with appealing qualities and discover that the remainder either do not meet your company's needs, or are not interested in the opportunity. Do not be offended if a foundry that seemed like such a good candidate for your product immediately rejects it; this selection process takes time and effort, so a disinterested foundry will do you a favor by rejecting you early on. It is always a good idea, however, to ask them about the reasons why they rejected you, because there may be something to learn or perhaps a misunderstanding to correct.

The ideal short list will consist of 2–5 qualified foundry candidates. Of course, it is possible to have more on the list, but from this point onward, you will begin to expend a lot more resources and time in your evaluation, so it is best to be selective.

Send a Request for Quote (RFQ)

In order to begin a thorough evaluation of the short list foundry candidates, you must be prepared to tell each of them more detail about your MEMS product and your company's anticipated volume production needs. Our method is to gather all of this information into one document which we call a Request for Quote (RFQ). The RFQ will contain your confidential information, so prior to sending it, you should pursue a nondisclosure agreement (NDA) with the candidate foundries.

There is no industry standard format for the RFQ; however, at minimum, it must cover the following vital topics:

- Short summary of your company, its business history, and the intended market application(s) for the MEMS product.
- Preferred wafer size and expected annual volume of wafers to be ordered over the next 3 years.
- A simplified cross-section diagram of the MEMS device, an outline of its process flow (materials and thicknesses, not the actual process recipes), and the die size.
- Requirements for wafer dicing, inspection, and electrical wafer probe test (if any).
- A list of information you would like to receive in the foundry quote, such as:

  - Nonrecurring engineering (NRE) costs.
  - First batch cost and timeline.
  - Look-ahead cost estimate for production wafers or known good die.

It is important to present facts in the RFQ and also to state any unknowns or uncertainties. Avoid the temptation to exaggerate your potential wafer volume, timeline to volume, or market applications in order to look more impressive. The foundry needs accurate information to provide a reasonable assessment.

Send the RFQ to the respective sales teams at the short list foundries. Plan to have follow-up phone calls with each sales team in order to answer questions about your document and to provide more details.

If possible, visit each foundry and give a presentation which provides more information about your business and its go-to-market strategy, the MEMS technology, the longer term market opportunity and anticipated wafer production volume. Your presentation will provide vital business context to the foundry team as they assess the opportunity.

When foundries are busy and have limited remaining production capacity, they will be very selective on what work they will accept. Customers might believe that they are the ones choosing the foundry; but often, it works the other way around: the foundry chooses its customer. So it is very important to audition well and to help a foundry to understand why your company could be a good fit and a valuable customer. To be an ideal fit, your MEMS product should not require the foundry to develop any new process steps or to use new materials; use a straightforward, low-risk process flow that they already understand; offer good profit potential; and have certainty about future wafer order quantities.

In general, expect the foundry to take 2–4 weeks to respond to your RFQ. They need time to gather their engineering team to evaluate your MEMS process, to do some initial business diligence on your company, and to make their own assessment of the market opportunity for your product. If they do not find good alignment with their own capabilities and market strategy, the foundry may decline to provide a quote. As before, if you get rejected at this stage, do not be offended, but do follow-up with the foundry in order to better understand the reason for their rejection.

Your company's in-depth evaluation and diligence process will begin in earnest when the foundry quotes arrive. Every foundry writes quotes in its own style, so you will need some time to study and interpret each of the quotes in order to make an equitable "apples-to-apples" comparison among the foundries' offers.

Pay attention to how the foundry responded to your RFQ. You can find evidence of their expertise with your technology in how they structured the initial work plan and any risks they identified. Any foundry quote that clearly grasped the nuances or challenges of your technology is worth pursuing further.

Do not immediately dismiss a quote that appears to be unreasonably expensive. A high quote usually indicates that the foundry sees risk in your product, perhaps due to unfamiliarity with your technology or its market, or could be momentarily busy. It is worth spending the time to dig a little deeper and understand how they arrived at their numbers. The quote is an opening offer and not necessarily the final firm price.

Due Diligence

After receiving any quote, you must plan a visit to the foundry. The purpose will be twofold: to spend time discussing their quote in detail and to begin your due diligence. Visit the foundry even if you already know some of their personnel or have visited them in the past. Things change all the time and you must have up-to-date information.[1]

Plan for a minimum half-day meeting, or ideally, a full day. At least two people from your company should plan to attend this meeting: a technical expert who is very familiar with all aspects of the MEMS design and its process, and a business executive who understands the long-term needs of your company and who has decision-making authority.

Ask the foundry to have their likely project manager and any key technical staff available to attend the meeting. You should also have the opportunity to meet the executive team. Ideally, there should be some planned time to socialize with the foundry executives and key staff, perhaps over a meal. You are contemplating a long-term business relationship with these people, so you need to get to know them.

It is important to be well prepared for this meeting, just as you would be to meet with a future investor or business partner.

Some things you should do to prepare:

- First, hold a strategy meeting within your company in order to identify the most important technical and business requirements for your product's success. What are your "must haves" and "nice to haves"? What are your deal-breakers or "no-go" items?
- Do your homework on the foundry:

  - Read their website and any available documents.
  - Review their annual report, if available.
  - Search for press releases and other public documents to identify their current customers and the types of technologies being manufactured.
  - Search patent databases to better understand any foundry IP, especially if you might be incorporating that IP into your product.
  - Review their quality certifications.
  - Prepare a list of questions.

- Prepare a presentation about your company and its product(s):

  - Company profile and overview.
  - Product features and market strategy.
  - Introduction to the MEMS device.
  - Pricing and volume requirements.

---

[1] In 2012, the former MEMS Industry Group created the "MEMS Foundry Engagement Guide," an online how-to wiki, to which many MEMS colleagues, including the authors, contributed. This website contained many useful tips and guidelines on diligence and is unfortunately no longer available.

- You may also want to prepare a statement of work (SOW) document for discussion with the foundry. Eventually, the SOW would be formalized into a project execution plan. The document should address:

  - Goals and criteria for success.
  - Roles and responsibilities of all parties, including subcontractors.
  - Transfer of mask data and process flow information.
  - Project tasks.
  - Project timeline.
  - What will be delivered by the foundry.
  - Communication and documentation methods.

During the visit, plan to assess the people, the business, and the facility. In addition to your own instincts about people and their personalities, you should be evaluating:

- Are the foundry's engineers engaged, perceptive, and have good attention to detail?
- Whether they have a problematic "not invented here" attitude. If they are not open to your suggestions and feedback, or unwilling to collaborate or compromise in this first meeting, be very wary.
- Are they good communicators? This goes beyond language fluency. Can they clearly articulate complex issues?
- Are they comfortable with regular update calls and visits?
- Are they open to accommodating rush orders and special requests?

MEMS development is a long undertaking and stressful situations will undoubtedly arise. The foundry team should have a positive, cooperative chemistry with your team in order to be able to successfully address the inevitable setbacks.

When evaluating the foundry's business, you should seek to understand:

- Is the foundry seeking your business for strategic reasons, or is it just hungry for any new work?
- What is their market strategy for the next several years?
- What is their roadmap for keeping their tools and processes current?
- What is their track record for developing complex MEMS devices and ramping them up to volume manufacture?
- What are their quality methods and policies?
- Are they adequately capitalized and have access to a labor pool in case of growth?
- How stable has the executive team been?
- Is their mindset to promote symbiosis and growth of the business relationship, or is it more transactional?

When touring a facility, try to go inside the cleanroom, if the foundry will allow it. If the foundry will not allow outsiders within their facility (which is often the case), ask for a window tour.

Items to observe include:

- Treatment of work materials. If unattended wafers are laying around, wafer boxes have been left open, tools are spread around, this means your work product might be treated carelessly.

- Condition of equipment and facilities. Vintage of the tool is less important than whether it is up and running. Facilities should appear well maintained and organized. Large floor stains or water puddles in the equipment chases are signs of maintenance issues.
- MES and inventory system. Is every process step scheduled and logged? How are steps monitored and validated? What is the average step cycle time? What are the up-time percentages and utilization of the tools critical to your process?
- Number of personnel working in the fab, which can indicate how busy the foundry is (relevant only to 150 mm fabs). Note, 200 mm and 300 mm fabs are highly automated so there is not necessarily a correlation between number of people and fab utilization.
- Tool redundancy. For your product's key process steps, does the foundry have at least two tools which could perform the process?

Additional Diligence

Even if the first meeting and diligence are successful, it is worth taking the time to talk to any current or former customers of the foundry, as well as the foundry's suppliers. The most valuable references will be from companies that have developed a MEMS product similar in process or complexity to yours at the foundry. You should also pursue "back channel" references not provided by the foundry. Ask your industry colleagues if they know of any companies who have worked with that foundry and try to speak to those companies.

When you speak with a reference, a key question to ask is, "How did the foundry resolve a process failure or mystery?" Listen carefully to determine the scope and thoroughness of the foundry's approach to solve the problem; the testimonial will provide clues on the foundry's expertise and professionalism.

Most importantly, during a foundry evaluation process, do not rush ahead just to close the deal! All too often, it is easy to get impatient, or to succumb to pressure from executives or investors to make a decision and get on with it. When in a rush, it becomes easy to skip over diligence, to overlook tell-tale bad signs, or to rationalize that some troublesome issue could be mitigated with close management. Having to later switch foundries due to an ill-considered or impulsive choice could take at least a year and a million US dollars, a penalty large enough to kill a start-up company or to hamstring even an established company. So spend the extra month, or whatever time may be needed, to be as thorough as you can with your diligence.

Quality certifications by End Market

Depending on the end use market for the MEMS product, your company's customers may demand that vendors in your supply chain hold specific manufacturing quality certifications. Verify that the foundry has the appropriate quality certifications for your end market, has a compliance plan, and has plans to keep renewing the quality certification (Table 18.1).

Table 18.1 Quality systems required by different end use markets

Market	Quality certificate for manufacturing
General	ISO-9001
Aerospace	AS-9100
Medical	ISO-13485
Automotive	IATF-16949

Make sure that at least one person on your team is familiar with the specifics of any quality systems required by your customers and can evaluate and later audit the foundry for compliance.

Negotiating the Contract

Work begins with a foundry at the start of advanced prototyping or early production, when needs and dollars at stake are relatively low. The foundry contract, however, may continue to govern as the complexity of the engagement and dollars involved increase over time. Needs and risks will further evolve as the foundry approaches readiness to do high-volume manufacturing and your company gets ready to sell product. It is therefore very important to spend time early on in the foundry contract negotiation to consider your longer range business needs and risks. It is never too early to negotiate for the terms that you know will be important to you in later stages of development and manufacturing.

Engage a business attorney familiar with manufacturing contracts to review the foundry's contract and to negotiate and draft your desired terms. If the contract will be governed by the laws of a foreign country, consider hiring an attorney familiar with that nation's laws and business culture. The foundry contract will govern millions of US dollars' worth of work, so it deserves careful attention.

There are several key business concepts and practices, described below, which are unique to MEMS foundry relationships and should be considered carefully prior to entering a negotiation. Typical and generic business contract terms like warranty, limitation of liability, etc. are not discussed here.

IP Indemnification

If the foundry will utilize its own process IP or incorporate its own device IP (for example, through silicon vias) in order to fabricate your MEMS product, then ideally, the foundry should fully indemnify your company against any patent infringement claim against their IP. Suitable remedies to resolve an infringement claim would be for the foundry to fix the item so that it no longer infringes the third party's IP; pay for license(s) to the patents being infringed; and/or compensate your company for additional design, testing, and qualification work which might be needed to create and qualify an alternate option based on non-infringing IP.

It might be tempting to minimize concerns about process IP infringement, based on the fact that a foundry has already been using its process for a while without any conflicts. However, we have seen MEMS patents issue which are, in our opinion, obvious, or have already been documented in the public literature, or have already been claimed in other issued, active patents. In the low probability scenario where a troll or other patent owner might try to assert their patent rights with an infringement claim or injunction against your company, you would certainly want the foundry to help defend any actions resulting from use of its IP.

Second Source Options

At some point in your business operations, it may become necessary for your company to engage a second foundry for wafer fabrication. There are many reasons to engage a second source foundry: security of supply, in case of fab fire or other disaster; to protect against unforeseen changes in the foundry's business which could, for example, cause them to unexpectedly increase prices or switch to a larger wafer size which might not make economic sense for your company; for the option to manufacture the wafers within a country where the bulk of your product volume is being assembled and sold, and so on.

If your product depends on the first foundry's process IP, enabling a second source for production would require obtaining a license from the first foundry to the process IP and possibly negotiating a royalty agreement or other compensation. If possible, also securing the first foundry's assistance to transfer the process would dramatically lower your risk, cost, and time of bringing up the process in a second facility. This would be an ideal outcome.

In reality, foundries are typically very reluctant to allow their processes to be transferred to other foundries. A foundry invests heavily to perfect a process, often absorbing expenses along the way, and they recoup that investment by manufacturing wafers made by that process. Proprietary processes are viewed as competitive advantages to offer to customers, with the additional benefit of keeping those customers bound to their facility. If a foundry might be willing to allow their process to be moved to a second foundry, they will likely be unwilling to assist in the transfer. Most foundries are motivated to create a high-energy barrier to prevent customer departures.

Some considerations for negotiating on second sources:

- If a foundry operates multiple facilities in different geographic locations, request that they guarantee that the MEMS product could be made in more than one of their facilities, so that if the main fab goes down (such as due to fire), there is a second facility to continue uninterrupted production.
- Request the right to transfer the process to a second source after x years of production, where x is greater than 3 years. To make the foundry feel even more comfortable, you might propose a condition whereby you would not transfer the process to any direct competitor(s) of the foundry.
- Offer a royalty payment in exchange for having the right to transfer the foundry's process to another foundry. As long as the process is not the foundry's "crown jewel," it is possible they would be open to this offer, at the right price.

Termination and End of Life Order Rights

As time passes, a foundry's management might change, its business strategy might change, and/or its equipment set could change (e.g., to larger wafer sizes). Any of these situations might cause the foundry or your company to decide that it no longer makes sense to do business together.

In the contract, request at least 12 months' notice for termination by the foundry or for any situation which might affect your company's ability to procure wafers, such as fab renovation for a larger wafer size. You should also request the right, triggered by specific events such as termination by the foundry, to a guaranteed product "end of life" order. The "end of life" order quantity should be large enough to fulfill at least 2 years' worth of your company's needed inventory. In other words, it should cover the length of time it might take to bring up and stabilize your product manufacturing at a second foundry. This "end of life" purchase should be offered at then-current pricing.

Exclusivity

If your company would like to request any exclusivity, the time to negotiate for it is at the beginning of the relationship. Foundries are usually reluctant to grant exclusivity, because working on multiple similar devices for different customers is another way they recoup the costs of tools and process development. Exclusivity might be granted for very narrow market definitions, or to exclude a limited list of competitors, in exchange for compensation. Some foundries will allow an exclusivity period, such as 5 or 10 years, after which they can work with whomever they prefer.

Acceptance Criteria

After the foundry has qualified your company's process for volume production, the fee structure should migrate away from NRE fees to a pay per wafer or pay per known good die arrangement. A key part of this business arrangement is defining the characteristics of a "good" wafer or a KGD which would merit payment. These essential characteristics should be set forth in a list of "acceptance criteria" which may be formally appended as an exhibit to an existing contract.

The acceptance criteria should clearly define the types and conditions of electrical probe tests and/or optical inspections to be performed at end of wafer process; the number of samples per wafer or lot to be tested; and a range of acceptable, or "pass" values for each key test item. Note that a foundry will only agree to criteria for parameters that they can control and have the means to measure.

Acceptance criteria should be based on well-understood and data-driven requirements for your product, not on theoretical or idealistic values. Acceptance criteria which are unnecessarily strict will result in rejection of otherwise useable material, lower wafer yields, and ultimately increase costs.

The contract should also describe what will happen if a finished group of wafers do not meet the acceptance criteria.

Export Controls

Many MEMS devices are strategic technologies that confer national competitive capabilities and therefore may be subject to export control, whether or not they might be used for military applications. Depending on the location of a foundry, receiving your company's MEMS designs and then shipping partially or completed MEMS products may require export licenses. For some types of MEMS technologies and some countries, export may be prohibited entirely.

For example, in the United States, MEMS technologies such as gyroscopes and infrared bolometer detectors are, at minimum, subject to export control under the Export Administration Regulations (EAR), set by the US Department of Commerce. Especially high-performing MEMS devices intended for military applications will be subject to additional control according to the International Trade in Arms Regulations (ITAR), set by the US Department of State. The EAR Controlled Commerce List (CCL) is much broader in scope and regulates many more technologies than the ITAR Munitions List, because it also regulates technologies of national strategic importance and dual-use technologies (useful in both civilian and military applications). Violation of either the EAR or ITAR regulations could result in large fines and criminal prosecution within the United States.

A MEMS product company should engage expert assistance to review the export regulations in all countries in which it plans to operate, fabricate, and sell product. When considering foundries located in foreign countries for production of export-controlled items, a MEMS company must review its risks and requirements prior to finalizing a contract with the foreign foundry. Because every country's export regulations and lists of controlled technologies will change from year to year, compliance with export regulations must be actively managed.

Select a Foundry Based on Best Fit, Not Lowest Price

Ultimately, it is crucial to select the foundry which offers the best overall fit to your company's business and technical needs.

While it might be tempting to select the foundry that offered the lowest bid, keep in mind that the initial quote is based only on the near-term work plan, and any forward-looking pricing is based only on the information that is available today.

Because the MEMS process and work plan will evolve as you pass through the stages of development, the initial foundry quote actually provides very little information on what the overall development costs are likely to be!

The lowest total cost will occur at the foundry with which you have the best compatibility. By the time you finish due diligence and negotiating with the foundry on the legal contract, you should have a very good understanding of compatibility. When technical expertise, business model, and management culture between the foundry and its customer are well aligned, the development will proceed as efficiently as possible. Over the long term, that synergy will result in the lowest cumulative cost.

Signing the contract and issuing your first purchase order to the foundry is just the beginning of this significant long-term relationship. In Chap. 19, we present best practices for efficiently transferring your MEMS product technology from the R&D facility to your new foundry partner.

Summary

A well-matched foundry is a strategic business partner that will contribute to your company's success by accelerating product development, enhancing product performance with process technology, and ultimately keep your product flowing to the marketplace.

Selecting a foundry is a major business decision for your company that deserves time and close attention from key technical and executive staff. It is a crucial step in MEMS product development that will consume at least a year and a million US dollars if done well; if done poorly, it could cause major delays to product launch and even bankrupt your company.

References

1. Fitzgerald AM, Jackson KM, Chung CC, White CD (2018) Translational engineering: best practices in developing MEMS for volume manufacturing. Sens Mater 30(4):779–798
2. Van der Waal R (2018) Choose your MEMS partner. Philips Innovation Services. https://www.innovationservices.philips.com/app/uploads/2018/05/How-to-choose-a-MEMS-Foundry.pdf
3. Fitzgerald AM (2015) How to successfully transfer MEMS from a university lab to a commercial foundry. Presented at Transducers 2015, Anchorage, AK 23 June 2015. https://www.amfitzgerald.com/s/150623_Transducers15_AMFitzgerald.pdf

Chapter 19
Transferring Technology for Production

During the foundry evaluation and selection process, the foundry received only enough information about your company's technology to assess the business opportunity and to provide a quote for the initial feasibility batch. They did not receive enough information for actually making your product. After the contract has been signed between your company and the foundry, and a purchase order issued, the transfer of your company's detailed design and fabrication information to the foundry team must begin, a period known as "tech transfer." This is the period in which your company will teach the foundry what you have made and learned so far, so that they can begin to set up their equipment and processes in order to replicate your product in higher volumes.

Preparing for Tech Transfer

When it is time to begin tech transfer to the foundry, you will first need to provide the foundry team with very detailed information on your technology. Review Chap. 16 for a list of all of the documentation you should already have on your product technology; if you did not have it previously ready, now is the time to do it.

Determining What Information to Share

When preparing information for transfer to volume manufacturing, it is helpful to have a strategy and to be organized. If you were to just dump a bunch of data on the foundry, they might eventually work their way through it, but you would pay for their effort in both time and money. Focus first on the process flow of record and mask layout (see next section) that you would like to transfer to the foundry and build up the supporting documentation from there, sharing more details on process steps that affect critical features of your device. Also, listen to the questions the

A. M. Fitzgerald et al., *MEMS Product Development*, Microsystems and Nanosystems, https://doi.org/10.1007/978-3-030-61709-7_19

foundry team asks regarding the process flow. Those questions likely reflect where they might be less experienced or where they have learned that those details are the difference between success and failure.

There are also trade-offs between compartmentalizing information to protect IP versus sharing information to engage partners in efficient problem solving. Ideally, concerns regarding data protection, potential conflicts of interest (such as the foundry also making your competitor's product, a common situation), and business strategy (pure play vs. mixed use foundries) were addressed during the foundry selection process. In general, we recommend that you not be too paranoid. If you know a special trick for getting just the right flavor of oxynitride from a PECVD tool that will improve your device's yield or performance, share it with the foundry team. Foundries know lots of processes and see many MEMS designs. When it comes to wafer fabrication processes, they probably already know the recipes you might think of as "secrets" and withholding information essential to your device's success will just slow down progress.

If you are trying to compartmentalize information to protect very sensitive product IP, avoid sharing the back-end process information, test data and analysis, and so on, with the foundry and disclose only those details relevant to the foundry's work. For example, if a special anti-stiction coating helps your device to perform better, and your company will apply it to finished wafers, not the foundry, then do not put details about that coating in your process documentation. Instead, disclose only your requirements, for example, the wafer must not have material X because it is incompatible with your coating process. Creating and enforcing some simple boundaries can effectively protect sensitive or trade secret information without jeopardizing your overall fabrication process.

Process of Record (POR)

The POR provided to the foundry needs to be much more detailed than what was provided during the foundry quote and selection process. The POR is not just a document, it is also a training tool for foundry engineers who are about to work on your product. Write it as if you were training someone new to run your process. (See Chap. 16 for more details.)

At a bare minimum, the POR that is transmitted to the foundry should include:

- An (x, y) mask layout with defined layers.
- A written process flow.
- List of critical process tolerances.
- A 2D cross-section (drawn to scale) for each process step.

Additional process information that can be helpful include:

- Examples of previous process designs and iterations, providing an overview of the history of how the design evolved to where it is today.
- What processes you had previously tried and what worked and did not work.

- 3D renderings of the process steps
- Sample die to the foundry engineers.
- SEM or microscope images of each process step from prior runs.

The foundry MEMS engineers will know how to interpret the information on your runsheet in order to set up their own process runsheet for their specific set of tools. Transferring the additional knowledge items listed above could save the foundry team from revisiting processes or design changes that you have already tried and discarded. Your engineering team should also be able to explain to the foundry's engineers why you have made certain process choices and to provide insight on which processes might be adjusted without harming device function versus which ones must remain as they are.

Planning

Before starting the tech transfer, it is a good time to revisit your company's go-to-market plans and product roadmap. The project timeline you set with the foundry must be informed by your company's needs and goals, as well as by any downstream back-end processes, assembly and tests that are needed to finish product fabrication and to prepare products for sampling or sale to customers. Telling the foundry you need wafers as soon as possible is not very informative; everyone says that. Instead, explaining why certain wafers are needed by certain dates and how those milestones link to your company's product ramp-up, and eventually, volume wafer orders, will get their attention and recruit their efforts to keep your wafers' process schedule on track.

Similarly, communicate within your company about realistic tech transfer timelines, which have been confirmed by the foundry, so that your executives and investors know what to expect. As you can see from Fig. 19.1, you are about to begin a foundry production development effort that will span at least 1–1.5 years.

Initial Transfer

In the semiconductor industry, which has the benefit of being several decades more mature than the MEMS industry, tech transfer occurs via software automation: automated design rule checking of photomask data, process rule checks, and data upload to the foundry. In the MEMS industry, in all but a few cases[1], tech transfer still happens the old-fashioned way: teams of engineers exchanging documents and talking

---

[1] These few cases usually involve established foundry platform processes, as mentioned in Chap. 11. If you are transferring a design and process to a very high-volume foundry such as TSMC or a large captive foundry, the technology transfer process may be formalized and you would need to follow their instructions. High-volume or captive MEMS foundries may also offer design rules or automated mask checks.

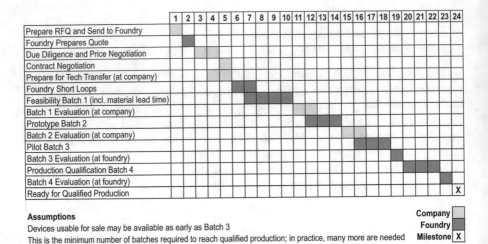

Fig. 19.1 Example Foundry Transfer and Production Ramp-up Timeline

about them. This process will go most smoothly if your documentation is complete and your company's process engineers are available for multiple meetings with the foundry's team, ideally with at least one being held in person. Even with a detailed POR to provide the foundry, if you were to simply "throw it over the fence" to the foundry, do not expect very good results. Tech transfer is far more involved than simply uploading documents. Your engineers should be prepared to meet with the foundry on a regular basis throughout the entire foundry feasibility stage, and beyond, until production is reached. It can be extremely helpful to create and collectively agree upon with the foundry team a plan that outlines the transfer and first processing run, what needs to be done and by when, where each step will take place, and who will be responsible.

Translating an Advanced Prototype Runsheet for a Different Toolset

Once the foundry is up to speed on your POR, the next step will be to work with the foundry's engineers to translate your advanced prototype runsheet to their toolset. Your runsheet was developed on your development fab's set of tools, and it is highly unlikely the foundry has the exact same makes and models of equipment. The foundry engineers must therefore figure out a way to get the same films, stresses, material qualities, and features from their own set of tools. Even if the foundry happens to have the same model tools that you used in development, you should be prepared for the possibility that the tools might give slightly different results. Just because your process worked perfectly on your tool does not mean that the foundry will be able to duplicate the process on their tool the first time through. Understanding what films, materials, and processes could be substituted if the process cannot be exactly duplicated can also help with the transfer process.

A key part of the transfer and translation process is also developing the acceptance, or pass/fail, criteria for each step of the wafer fabrication process. Identify which key specifications are design- or process-based and establish how they will be confirmed. If there is no clear division of labor, then you may need to handle it or work with the foundry to do it together.

The foundry's operators eventually need a set of criteria against which to judge whether the process was correctly executed or not. At this stage, you must be quantitative and specific about what is essential in each process step. For example, it is not enough to specify the alignment between a piezoresistor element and the edge of a membrane in a silicon pressure sensor by the mask layout alone. You also need to provide an acceptable alignment tolerance (e.g., ± 3 μm). Another example for a pressure sensor would be the tolerance on the bridge resistance of the sensor. Specifying a tolerance of $\pm 1\%$ versus $\pm 10\%$ could be the difference between giving the foundry a difficult process requiring many short loops to stabilize versus one that is easily achieved. Process tolerance values may be based on your testing of advanced prototypes, or on prior modeling, and they all must be measurable by the foundry.

Providing the foundry with this specific process acceptance data will dramatically increase the odds of successful process transfer and of achieving working devices at the end of the foundry's feasibility stage. Keep in mind that at a foundry, the process engineers may not look at your wafer at every step and cannot make changes on the fly to adapt to unexpected results. If you feel anxious that your wafers really need to be closely monitored by a foundry process engineer, then your product may not be truly ready for foundry production (Review the readiness criteria in Chap. 17).

Wherever there might be missing information between process specifications and acceptance criteria, then short loops, wafer splits, or staging can be used effectively to discover that information. (Please refer to Chap. 15 for more information on these fabrication risk mitigation methods.) Identifying this vital information and taking time to discuss it with the foundry engineers will help you to capture maximum value for the money spent at the foundry and will set both parties up for a successful start to a long relationship.

Transferring Photomask Layout Data

The bare minimum mask documentation you would prepare for the foundry is a mask layout file and a list of all of the mask layers. Even the simplest MEMS mask layout design benefits from additional documentation describing key mask features, the alignment strategy, and the mask digitization information (data digitized dark or clear, polarity). Please see Chaps. 11 and 16 for some tips for managing and documenting the mask layout.

Automated design rule checks (DRC) for MEMS mask layouts usually do not happen with a custom fabrication process. If you happen to be using an established process or a foundry platform process, then your layout may be subject to the foundry's DRC. If you are using a platform process at a foundry, design rules should have been provided to you by the foundry. You will need to confirm that your mask layout conforms to them and to resolve any discrepancies before the foundry will accept your mask data for production.

Whether using a custom or a platform process, you will likely only need to deliver the unit cell of your device's layout, which the foundry's mask engineer will then use to populate a photomask reticle and/or wafer layout, in the case of contact lithography. Even with a well-defined and documented mask layout file, the data delivery process to the foundry can be more time-consuming than you might expect. You will need to learn the foundry's mask data submission process, which is different for each foundry. Then the foundry will need to add its own alignment marks and process control monitor test structures (PCM). Because of this added data, you may never be allowed to see the final mask file. It is now considered property of the foundry because it may contain the foundry's proprietary information.

The foundry will, however, send some form of the final mask layout as a "proof sheet" for you to double-check. This is a very important step! Do not blindly assume that the mask data has been correctly processed, no matter how many conversations you have had. You may need to go back and forth with the foundry's mask engineer a few times before the mask layout can be finalized. It can be especially helpful to create a document outlining the photomask data file transfer process and the responsible parties for each step (your company or the foundry) because so much information is being transferred back and forth. If a mask revision would ever be required at a later date, then this process will need to be repeated.

When the mask data transfer is finally complete, we recommend that you document what you sent to foundry and why, for your own internal records as well as to assist future photomask layout data submissions.

Monitoring the Foundry Feasibility Stage

In the foundry feasibility stage, the purpose of the wafer runs is for the foundry to translate and to replicate your advanced prototype process on its own tools. These first few runs are for learning and transfer of information, not about fabricating working product[2]. It is not unusual for wafers from these first runs to be only partially finished or to not yield any functional devices at all. If you find yourself in this situation, you can take a little comfort in knowing it is not the first time it has happened when transferring MEMS technology.

---

[2]Your company executives and investors, however, will likely be putting pressure on everyone to deliver working product out of these initial runs. Educating them on the foundry stages of development (Chap. 3) can help to adjust their expectations.

This can be a very high-pressure period of time for any company, but especially for start-ups, whose access to more funding may depend entirely on its success. While you are in this critical learning phase of the transfer, it is essential to stay in close touch with the foundry team, to increase the chances of a successful transfer. Some good practices include:

- Establish the main points of contact at the foundry and at your company.
- Set up biweekly meetings between the teams.
- Have the foundry provide a visual report on wafer progress, with pictures, data, etc. for at least the first wafer lot run. Reporting during later runs, once the process is stable, can become more administrative.
- Have a project manager and MEMS expert within your company on call, in case any issues arise which need prompt attention.

During the first runs, you will need to let the foundry do its job, but do not expect them to do all of the work to monitor the project results. Without regular input from you, they may not have the all information they need to successfully complete the first wafer runs.

The Foundry Engineers Are Your Partners Now

The foundry engineers are your expert collaborators now, and if you treat them as part of your development team, you will reap the most benefits from the relationship. The foundry team brings a large body of knowledge and experience from having transitioned other clients' products to high-volume manufacturing. They have a distinct skill set on developing processes for stable production, and know what is possible or not possible on their tools. The foundry team wants you to be successful, because their business model and long-term profitability depends on reaching and sustaining high-volume production of your product. They are not in this business to do endless research and development, so if the foundry team is suggesting short loops or other work, such as a design modification, it is because they are trying to reduce long-term fabrication risks and to improve product yield.

Conversely, they should be treating your team with respect, because your team is the expert on your novel product design. Your advanced prototyping experience can help inform the foundry on how to best mitigate processing risks (see Chap. 15). Any conflicts with the foundry team must be resolved early on; any serious or irreconcilable personality clashes might need to be resolved with reassignment of personnel.

If you happen to discover any gaps between what the foundry's sales team promised you and what the foundry process engineers state is possible, these must be resolved immediately. If you detect that the sales team has dramatically oversold the foundry's capability, for example, claiming expertise with a process that the foundry's engineers are treating more like a new experiment, immediately bring this to their attention and work to get it resolved. If the foundry cannot quickly close the gap in process expertise, consider cutting your losses while still early in the transfer and switch foundries right away.

During the transfer phase, your team should take the time to visit the foundry in person again and make the effort to develop a professional rapport with the foundry team. If you only have funds to visit once a year, go when you are working on the technology transfer and process documentation. You should be meeting with the foundry engineers and project manager assigned to your company, and if possible, meet the operators on the key tools for your process. (This may not be possible at high-volume foundries.)

"Golden Wafers" and Murphy's Law

Sometimes, early in development, a "golden wafer" appears. A golden wafer is one that has devices with the exact performance you were hoping for, and/or has exceptionally high yield. (Often, it looks beautiful and flawless, too. Be sure to take lots of pictures of it for your marketing team!) Once you see a golden wafer, it is only human nature that you will expect all future wafers to also be golden, and thereby set yourself up for some potentially serious disappointment.

A typical story that we have often seen in our business: a golden wafer appeared in the last advanced prototype batch at the development foundry, and as a result, everyone felt that, at last, it was the right moment to transition to a production foundry. Everyone's expectations were set high that the foundry would immediately be able to replicate many more wafers with the same golden results. Then the first run in the foundry feasibility stage does not yield any golden wafers; then neither does the second. By this time, the customer would often be quite impatient and even angry—where are the working devices?

A golden wafer that appears in development can be a mirage—the appearance that you have figured out the perfect process—when in reality, it results from a lucky set of circumstances which you do not quite know or understand. If you are lucky enough to see a golden wafer, treat it as merely an existence proof, evidence that it is *possible* to get great devices. It will take a lot of hard work by the foundry to then tune and replicate the exact set of process circumstances which consistently yields golden wafers. The foundry's methodology for development (summarized in Chap. 3) exists to systematically work through all the process variables, using methods such as design of experiments, 6-sigma, and response surface characterization, to create a stable process that will make golden wafers by the thousands. This takes time.

Murphy's law (if it can go wrong, it will) applies in MEMS at every stage of the product development. It never ceases to amaze us how even seemingly simple processes can go very wrong. All your hard work on the advanced prototypes and during the foundry transfer helps to minimize the risk that Murphy will appear, but keep an eye out for him anyway. If you did not know if your process would work well the first time, then the foundry will not either. Be prepared to shoulder some of the weight of problem solving with your foundry partner. Everyone feels better about working together to get through tough points. It sends the right message that you are indeed a team.

Process Monitoring

Process monitoring is the task of regular measurement and data tracking of the wafer process at the foundry. Examples include measuring furnace temperatures, film thicknesses, and device feature dimensions. The foundry is typically well aware of which parameters are best monitored in order to monitor the health of each tool and each process step. You, on the other hand, understand your device best. As a result, work together with the foundry to establish a process monitoring plan. This may include designing process control monitor test structures with the foundry that address processes that are critical for your device. During the first run of the foundry feasibility stage, you may be able to review the process monitoring data with the foundry and make sure that the process is properly monitored within the correct limits.

Your ability to review all process monitoring data from every run often depends on the size of the foundry. This request would add to the foundry's burden and might slow down wafer processing. Moreover, the foundry is the responsible party for proper function of its tools. It can be argued that if you were to get too involved, you could become a partially responsible party if anything were to go wrong. Instead of reviewing all process monitoring data, you might request that the data specific to your runs be saved so that it could be reviewed at a later time, if and when a problem occurs.

Once a new process has been established and stabilized, routine process monitoring data would continue to be watched by the foundry personnel, but it would not usually be reviewed by you. If and when a tool performance problem emerges, the foundry would likely react to it and solve it without informing you.

Moving on to Pilot Production

Once the foundry completes the feasibility stage runs successfully, it will have met its own internal quality criteria for moving on to the next stage, pilot production. When you have reached this stage, the technology transfer has been completed, and the responsibility for the process is in the hands of the foundry going forward.

During the foundry pilot production stage, and later during production, your engineering team's interaction with the foundry technical staff will become much less frequent, perhaps even shrinking to an annual review. A very stable, long running production process may eventually be managed only by purchasing personnel at your company, who interact with an account manager counterpart at the foundry.

Managing Ongoing Risks

Even while foundry production is proceeding smoothly, any number of internal or exogenous risks are always present. Here are a few which deserve your attention.

Cost and Availability of Materials

Although the element silicon is abundant in the Earth's crust, the number of factories on our planet that can purify silicon ore and form single crystal ingots is limited. During economic boom times in the semiconductor and solar power industries, demand for silicon goes up due to the fixed capacity for processing silicon, causing the price for silicon to go up as well.

These market dynamics were especially acute during a boom period in the solar panel industry, 2006–2010, during which the price of raw silicon rose more than tenfold as demand far outstripped global processing capacity. [1]

Yet another group of factories perform wafering, the process of slicing single crystal silicon ingots into wafers and polishing them. Many MEMS devices employ silicon-on-insulator (SOI) wafers, which are also commonly used in the semiconductor industry. Manufacturing SOI wafers requires special equipment and processes and only a limited number of factories on the planet can produce quality SOI wafers. During 2017–2018, when there was heavy demand for SOI wafers due to a boom in the semiconductor market, the lead time for SOI wafers increased dramatically. In our business, we experienced an increase in lead times for custom SOI wafers, from the typical 4–6 weeks to 6–12 months.

The semiconductor and solar power industries, behemoths compared to the MEMS industry, drive the supply and demand cycles for silicon wafers. A MEMS product company and its foundry partner should carefully track market changes in the cost and availability of the wafer types needed for manufacturing the MEMS product. Foundries will often build up wafer inventory to buffer against anticipated market changes. Increasing wafer inventory ties up capital, so thoughtful financial planning and analysis with inputs from both partners is needed to make solid strategic decisions.

Facility Damage

Any number of potential damage events can take a foundry offline for days to years or more. The most common foundry disruptions occur from minor incidents, such as flooding from a burst pipe or perhaps a sudden tool failure, which can disrupt production for hours to days, and rarely for weeks.

Fire is a more serious risk because even minor fires can result in long downtimes, weeks to months, due to the need for cleanup from smoke damage and particles. Recovery from a major fire, as occurred in a fab owned by GE Druck in the UK in 2014, can take over a year [2, 3].

Less frequent but potentially catastrophic risks, such as earthquake, typhoon, and particulates from wildfires or volcanoes, exist at many fab and vendor locations, particularly those located along the Pacific Rim. Foundries in developing countries may also be at risk for interruption or damage from intermittent power outages.

More recently, in 2020, the global pandemic of COVID-19 affected foundry performance worldwide. While most foundries kept operating as essential businesses, the need for "social distancing" within the facility reduced the number of people who could be working at one time, thereby reducing fab capacity and wafer throughput.

MEMS product companies should continually evaluate their entire supply chain, not just their foundry partner, for robustness and then make sure that appropriate redundancies or back-up manufacturing plans exist. For products such as critical medical equipment components, it would be wise to identify second source vendors in a completely different geographic location in order to ensure business continuity in the event of a regional disruption.

International Business Issues

In addition to potential language barriers and working across different time zones, doing business with international suppliers carries a few risks that can be safely managed if handled attentively. One common risk is foreign currency fluctuation, especially for vendor jobs which might span months or years between initial quote to final payment. Fortunately, many vendors in the semiconductor and MEMS industries are accustomed to doing business internationally and are usually willing to quote and accept payment in your company's home currency in order to eliminate your risk of price change due to currency fluctuation.

Changing political positions on international trade agreements, import tariffs, and worker visas also have significant effects on the practicality and cost of doing business with a foreign vendor. Any change in import/export duties or new tariffs could make a once economical cross-border business totally impractical. Companies doing significant amounts of international outsourcing should consider adding domestic or third country vendors to their supply chain in order to hedge against sudden, unfavorable changes in trade policies.

Actively Manage Foundry and Vendor Relationships for Best Long-Term Results

Manufacturing a MEMS product is a complex undertaking involving hundreds to thousands of people, advanced technology and facilities, multiple disciplines, and a supply chain spanning the globe. Business and operations leaders, even those within vertically integrated companies, need to devote a portion of their attention to monitoring and nurturing the complex ecosystem upon which their company's production and success depends.

First and foremost, successful supply chain management requires tending to the human relationships, which are so easily and often overlooked. Until we reach the point where robots run factories and an artificial intelligence can make nuanced judgment calls, the strength of a company's supply chain ultimately depends on the strength of the human relationships between the entities. Investing time in getting to know the leaders of key supply companies, and in understanding their business's challenges and outlook, will help both parties when the inevitable problem or interruption occurs. Trust between people remains an incredibly valuable asset in the technology business.

It is also important to share business information and outlook, for mutual advance planning. Practically speaking, the foundry, and all the vendors in your supply chain, are your company's business partners. The foundry, to a large extent, is an investor in your company. Treat them with the same respect and openness as you would your other business partners or investors. You are in it together for the long haul.

Summary

Preparing for tech transfer starts with clear, concise documentation of your MEMS device layout and process flow. Ideally, your engineering team and the foundry team will meet in person to review the documentation and answer questions in real time. When the foundry has transferred the layout and process flow to their tool set, you will need to review their file for accuracy—do not assume that your data has been interpreted correctly. During the feasibility stage, meet with the foundry on a regular basis to discuss progress and issues as they arise. The foundry, with your input, will determine when the project is ready to move onto pilot production. These two stages can take 1–2 years. With your diligent preparation and with attention by the foundry, you can create the best conditions for the foundry production to run smoothly.

References

1. Mufson S (2013) Prices flat in polysilicon market graphic in Chinese tariffs may hurt U.S. makers of solar cells' raw material. *Bloomberg News, The Washington Post.* https://www.washingtonpost.com/business/economy/prices-flat-in-polysilicon-market/2013/07/23/914479d0-f3e4-11e2-9434-60440856fadf_graphic.html. Accessed 5 Aug 2020
2. Fire Industry Association (2014) Chemical reaction starts fire in Groby factory. https://www.fia.uk.com/news/chemical-reaction-starts-fire-in-groby-factory.html. Accessed 5 Aug 2020
3. Hughes B (2015) GE Measurement & Control Opens World Class Silicon Clean Room in Groby, UK. Press release. https://www.bakerhughesds.com/news/ge-measurement-control-opens-world-class-silicon-clean-room-groby-uk. Accessed 5 Aug 2020

Chapter 20
Manufacturing Test: Opportunity, Cost, and Managing Risk

In this chapter, we define manufacturing test broadly as the practice of monitoring the entire manufacturing process, and identifying and removing defective products to ensure the customer will receive good products. It begins with measuring and testing incoming raw materials and ends with final checks before the product leaves the manufacturing site.

Manufacturing test could be regarded as the "nervous system" of the MEMS production process, continuously gathering data at each step of the process. This data allows the MEMS company to monitor the health of the production line and to improve the MEMS product.

Manufacturing test differs from development test in that it focuses solely on identifying defective parts. Development test, in contrast, focuses on device performance characterization, product qualification, and failure analysis. Table 20.1, introduced in Chap. 14, reminds the reader of some of the high-level differences between development test and manufacturing test.

Opportunities in Manufacturing Test

Manufacturing test offers multiple benefits to a MEMS product company, including:

- Improving the quality of the MEMS product.
- Increasing the MEMS product market value.
- Reducing time to product introduction and revenue generation.
- Improving yields and lowering product cost.
- Differentiating products.
- Improving supply chain efficiency.

The benefits are multifaceted and intersect with many aspects of a MEMS company's business. We discuss each of these benefits below.

© The Author(s), under exclusive license to Springer Nature Switzerland AG 2021
A. M. Fitzgerald et al., *MEMS Product Development*, Microsystems
and Nanosystems, https://doi.org/10.1007/978-3-030-61709-7_20

Table 20.1 Overview of differences between development and manufacturing test

	Development test	Manufacturing test
Test volume	Few devices	Many or all devices
Test data	Thorough characterization	Minimal characterization
Test time	Minutes to days	Seconds

Improving the Quality of the MEMS Product

The primary purpose of manufacturing test is to identify and remove defective products in the manufacturing stream so that the customer only receives good products. It is a crucial part of the overall product quality control. A single defective part costs money, time, and goodwill between the supplier and customer. If your customer is an integrator and assembles a defective MEMS component from your company into their own product, their product will in turn be defective. Your customer will then lose time and money.

If many defective MEMS components are unknowingly assembled into your customer's products, then your customer may demand compensation for their lost inventory, whose value is usually many times more than the MEMS component. An even worse situation would arise if your customer's now defective product were unwittingly sold, and an end user discovers the defect during in-field use, leading to warranty returns, loss of goodwill, and perhaps even lawsuits.

Thoughtful and effective quality control can prevent many problems that can seriously damage a business, which is why it has become a professional discipline unto itself, with its own standards and regulatory organizations. For start-ups and young companies, we encourage you to think about and implement quality control as early as possible in MEMS development.

Increasing the MEMS Product's Value

For MEMS products, manufacturing test offers opportunities to increase the product's value, in particular for sensors and actuators, by calibrating and compensating them to achieve higher accuracy and performance. Manufacturing variations create variation in a sensor's response to stimuli, such as offsets in bridge resistance for pressure sensors. This offset can be measured and compared to a calibration standard. After measuring the offset difference between the sensor and the standard, the sensor can be compensated via analog or digital circuitry or software, so that the sensor output conforms to the calibration standard as closely as possible. Improvement in a sensor's accuracy enhances the product's performance, and therefore increases its value. We discuss calibration and compensation in more detail later in this chapter.

Reducing Time to Product Introduction and Revenue Generation

When a manufacturing process is newly established and starting to output products, the defect rate is usually high. Depending on the complexity of the manufacturing process, it could take months, perhaps years, to bring the defect rate to down to acceptable levels.

Fortunately, your company does not need to wait until that point to ship products to customers. Manufacturing test can filter out defective products and enable the MEMS company to ship 100% good products to customers. While manufacturing test identifies good products, the rest of the development and manufacturing team can work to improve yield.

By being able to ship product sooner, effective manufacturing test critically enables the MEMS company to receive revenue earlier. This is an important turning point in a MEMS development project, when the long years of investment finally deliver revenue. For a start-up company, achieving revenue months sooner can be the difference between survival and bankruptcy.

Improving Yields and Lowering Product Cost

Test data enables your team to focus resources on the steps that need the most improvement in order to quickly improve overall manufacturing yield and, ultimately, to lower per-unit costs. As an example, the chart in Fig. 20.1 shows the yield loss at each individual step of a 50-step manufacturing process. Three steps in particular show higher yield losses. With this information, the MEMS product company could prioritize those three steps to improve product yield.

Every device that fails manufacturing test effectively increases the cost of the remaining devices. In a simple example, imagine that it costs US $1 to produce one MEMS sensor, and costs scale linearly with quantity, so 100 units are produced at a cost of US $100. After manufacturing test, only 90 of the 100 units passed. Because the US $100 production cost is a sunk cost, the per unit cost for the 90 remaining units is $1.11 per unit (US $100 for 90 sellable units).

Moreover, per-unit cost increases nonlinearly with increasing yield loss, described in Eq. 20.1 and shown in Fig. 20.2:

$$\text{Per-unit cost} = \text{Manufacturing cost} / (1 - \text{yield loss}\%) \qquad (20.1)$$

A 1% yield loss results in an approximately 1% increase in effective cost, but a 20% loss results in a 25% increase, and a 50% yield loss results in a 100% increase in per-unit cost.

Because manufacturing cost accumulates with every step of the production process, another way that manufacturing test could help to decrease overall per-unit

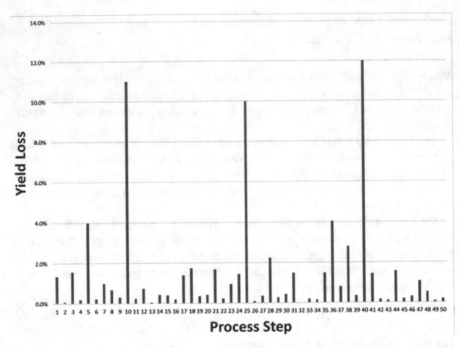

Fig. 20.1 Illustrative example: yield loss at each manufacturing step for a hypothetical 50-step manufacturing process. Three steps in this chart (10, 25, 40) have unusually high yield loss. Those steps receive priority for manufacturing process improvement

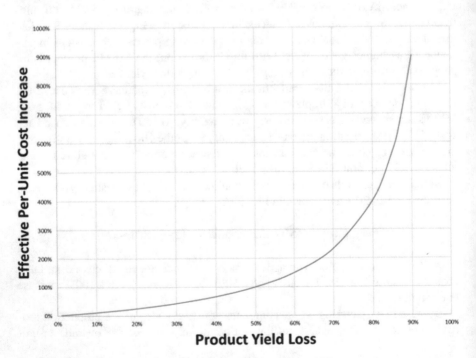

Fig. 20.2 Effective per-unit cost increases non-linearly with yield loss

cost is by preventing investment in processing a defective component. Extending the example above, suppose that a manufacturing test identifies the 10 defective parts when the sunk production cost is only US $0.10 instead of the final cost of US $1. The remaining devices that reach the end of the line are good, resulting in 90 good finished products. Now the total cost to manufacture the 90 good products is only US $91 = $90 + ($0.10 × 10), and the per-unit cost of each good product is US $1.01 (= $91/90 units).

A good example of this benefit is wafer probe. Wafer probe occurs at the beginning of the back end process (discussed in Chap. 12). The back end accounts for a significant part of overall MEMS product per-unit cost. Wafer probe can identify the good die on the wafer, and remove the bad die before going through the back end, thereby effectively decreasing the overall per-unit cost of a MEMS product.

Differentiating Products

A MEMS product can be differentiated for various potential customers by implementing different manufacturing test protocols. For example, an OEM consumer electronics customer may have looser sensor performance requirements than an OEM biomedical customer. The consumer electronics OEM's product may have an expected lifetime of 1 year and per-unit cost is a critical product requirement. The biomedical OEM's product may be surgically implanted into a patient and expected to function for at least 10 years. Product reliability is the biomedical OEM's major product requirement.

A MEMS company might sell the same MEMS product with the same design and same fabrication process to both customers, but the biomedical product would receive significant additional testing due to its stringent reliability requirements. For example, testing for the biomedical OEM may include hermeticity testing, 100% sample testing for quality, and more extensive test documentation. In contrast, the consumer electronics OEM wants to reduce costs and is willing to accept a more relaxed testing protocol, such as elimination of hermiticity testing, 1% sample testing for quality, and minimal documentation requirements. These variances in test protocols differentiate the products and impact their cost and value.

Another way that manufacture test can differentiate products is through a practice from the semiconductor industry known as "binning." MEMS device performance will vary, and this variation can, in some cases, turn a yield problem into a beneficial market segmentation. For example, sensors of a single design, from the same manufacturing line, may exhibit accuracies that range in a Gaussian distribution from part per 1000 to part per 10,000. Manufacturing test can sort the sensors by their accuracies into interval bins. Rather than throwing away the low accuracy sensors, they can instead be sold at a lower price. Likewise, the higher accuracy sensors can be sold at a higher price.

Binning can also apply to the consumer electronics vs. biomedical OEM example above. Devices that fail the biomedical OEM's test protocol may still be acceptable for meeting the consumer electronics customer's requirements.

Improving Supply Chain Efficiency

Manufacturing test can improve supply chain efficiency and reduce risk by compartmentalizing defects among the vendors participating in the manufacturing process. As each vendor completes its task, manufacturing test checks the quality of that work before the product goes to the next vendor. When a yield issue is discovered, it can be assigned to and addressed by the responsible vendor.

For example, a back-end-of-line (BEOL) vendor receives wafers from a foundry and packages from a supplier. Before the BEOL vendor starts work, the wafers and the packages are inspected. If, for example, the wafers are found to be defective, then the wafers may be returned to the foundry and reprocessed. This action contains the defective wafers before they would be further processed by the back-end vendor, and assigns responsibility of corrective action to the appropriate party.

Challenges of Manufacturing Test

The benefits of manufacturing test are broad, varied, and numerous, and its challenges are as well. Establishing this monitoring system to oversee the manufacturing process is a major investment which requires people, equipment, money, and time.

Manufacturing test requires its own full-time team, even for low volume production. The team's responsibilities include establishing and maintaining test capability, running tests, and monitoring and documenting all results. Members of this team will include mechanical, electrical, and software engineers, technicians, quality assurance personnel, procurement staff, and others.

The equipment needed in manufacturing test requires significant investment. To give a sense of the scale of the challenge, consider an example of a company using a single machine to test 24 hours per day, 7 days per week, 365 days per year. If the company wants to manufacture one million units per year, then each unit would have a maximum duration of 31 seconds to be tested.

There are many ways to approach this challenge. For example, this example assumes that each unit would be tested sequentially. The allowable test time could be increased by using multiple copies of test equipment, batch testing, pipelining,[1] and so on.

Test equipment is often customized to meet the varied requirements of each MEMS device and application. The function and environmental test requirements for a pressure sensor in a tire pressure monitor are quite different from an accelerometer in a smartphone. High-throughput testing equipment is further specialized by package size, shape, and mass in order to achieve its high speeds.

---

[1] "Pipelining" is the practice of distributing the tests so that they are conducted in parallel, but at different points of the manufacturing process

As a result, the test equipment likely will need to be developed, owned, and maintained by your MEMS product company, which is a major investment. The order of magnitude cost for such equipment is several hundreds of thousands to a few million US dollars per tool. The cost varies depending on several factors, including the testing throughput as well as the specific tests that the equipment must conduct. If you need to have more than one machine to meet your test throughput requirements, costs will quickly multiply.

Custom test equipment may need to be developed in-house or subcontracted to a development group. Like the MEMS device itself, the equipment will require its own design, build, and test iterations in order to be ready for use. The time needed for development of custom test equipment can be a year or more. If your company needs its manufacturing test equipment to be ready as soon as the first qualified MEMS products are available, test equipment development must be done in parallel with the product development. This means that testing requirements must be identified no later than the MEMS advanced prototypes stage.

Stages of Manufacturing Test for MEMS Products

MEMS component manufacturing test generally has four stages, in chronological order of testing:

1. Process recipe monitoring: monitors each step of the wafer fabrication process at the fab through observations at the wafer-level and process tool data. Typically done by the foundry.
2. Process control monitoring: uses test structures on the wafer, called process control monitors (PCM), that are electrically tested to evaluate the condition of the MEMS wafer at critical points in the process. Typically done by the foundry.
3. Wafer probing: electrical (and possibly thermal) test of some or all MEMS die on the wafer after fabrication. Typically executed by the foundry or the back-end vendor to the MEMS company's specifications.
4. Package-level testing: test, calibration, and compensation of each device after wafer singulation and die packaging. Typically executed by the MEMS product company.

Note that companies selling products incorporating MEMS components into boards or other complex system products will need testing appropriate to the level and type of product integration (not discussed here).

Process Recipe Monitoring

Process recipe monitoring watches and checks the wafer fabrication during every step at the foundry. Examples include measuring furnace temperatures, film thicknesses, and feature dimensions. This work is typically done by the foundry. The foundry knows which parameters are best monitored in order to confirm the health of each process step, but on occasion, may require input from your company regarding the specific requirements for your MEMS product. As discussed in Chap. 19, the foundry understands their wafer processing equipment best and you understand the MEMS device and its application best. Together, you and the foundry will establish the process monitoring plan that is best for your product.

Once that plan has been defined, you and the foundry might consider reviewing the process data from the first feasibility fabrication runs to ensure that the process is sufficiently monitored and tuned for your product's needs. The foundry may consider some of this data to be proprietary but they should confirm the data is being collected and within normal operating ranges. This is also the stage where the limits of the process window start being established.

Once the process is established and stable, routine process monitoring data continues to be watched by the foundry, but it is not reviewed externally by you, the foundry customer. If and when a tool issue emerges during the foundry production stage, it is the foundry's responsibility to respond and correct it internally, usually without informing you, unless the production schedule or wafer output becomes affected.

Process Control Monitoring

Process control monitors (PCM) are test structures used to check the wafer conditions at the earliest possible point at which wafers can be electrically probed, usually after the first metallization step. Examples of diagnostic tests include checking for electrical opens and shorts and measuring static capacitances and resistances.

Foundries have their own established PCM designs for common MEMS fabrication processes. For novel devices using a new fabrication process method or material, a PCM may not be available and would need to be developed.

The purpose of the PCM is not to determine whether each individual device on the wafer is good. Rather, PCMs determine whether the intended process has been executed correctly. If the wafer passes its PCM tests, then it continues fabrication. If it fails, then it is removed from the process flow to save further manufacturing investment into that wafer and may also trigger an investigation into the cause of process failure.

After wafer fabrication is complete, the PCM data can be evaluated against acceptance criteria to determine whether the completed wafer meets specification. The PCM data may be used to define the responsibility boundary between your company and the foundry. Wafers that fail PCM tests are usually the responsibility of the foundry. Wafers that pass the PCM tests (and any additional contractually obligated tests) would then be sold to your company. Once wafers have passed PCM testing, the foundry has fulfilled its obligation to manufacture the wafers and further risk for the wafers is now undertaken by you, the foundry customer. For novel processes or devices without an established suite of PCM tests, the wafer-level electrical probe results (discussed in the next section) may serve to fulfill acceptance criteria.

It is important to understand what PCMs will be used, how they will be measured, and the foundry's criteria for interpreting that data as pass versus fail. As discussed in Chap. 19, your company should collaborate with the foundry regarding how best to monitor the process execution in order to deliver best results.

In addition to the foundry's PCMs, we recommend that you establish your own set of PCM test structures on the mask layout. The foundry's PCMs are primarily intended for in situ process monitoring, not for device performance monitoring. The foundry PCMs are usually placed in the dicing streets of the wafer in order to save silicon area and are therefore destroyed during wafer dicing.

Your PCM test structures will be useful for understanding how process conditions influence MEMS device performance, particularly during the advanced prototypes and foundry feasibility stages of development. It is possible that foundry PCM data might look fine, indicating a correctly executed process flow, but the MEMS devices do not function quite as expected. This situation may occur when some underlying detail of the MEMS design was not well understood.

For example, if the influence of thin film stress on device performance had been underestimated by the designers, the MEMS company could have specified a broader range of allowable stress values to the foundry than would be suitable for good device performance and yield. So even though the foundry properly achieved the specified stress, the device still does not yield as hoped. Inclusion of film stress test structures on the mask, such as those developed by NIST [1], would be helpful to diagnose and solve this particular problem.

Wafer Probing

Wafer probe, also known as wafer sort, evaluates the wafer at a finer detail level than the PCMs. Wafer probe investigates the health of individual devices on the wafer, and typically every device on the wafer is probed and assessed pass or fail according to predetermined criteria. The good and bad die may be electronically mapped, an example shown in Fig. 20.3, or bad die may be marked with ink to visually distinguish them from good die. After the wafer is singulated, bad die are removed from the manufacturing process and may undergo failure analysis. The good die continue through assembly, packaging, and package test. These steps are part of the back-end-of-line (BEOL) process, which is discussed in more detail in Chap. 12.

The dataset from wafer probing is rich. Because there can be hundreds to tens of thousands of devices on each wafer, the statistical power of the data is high, and subtle effects can be distinguished. In addition, the data can be mapped to wafer location in order to view spatial trends. The patterns on the wafer map are a powerful tool to identify and address yield losses in the fabrication process.

However, while the dataset is rich, wafer probing usually cannot completely test your MEMS product, particularly if it is a sensor or actuator. Wafer probing is

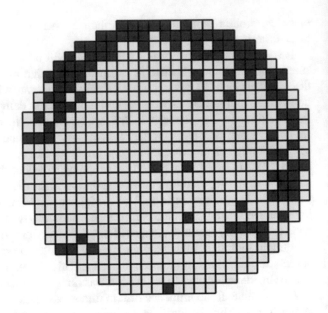

Fig. 20.3 Illustration of a wafer map after wafer probing. Good die which passed test criteria are indicated by light-colored squares, and the failing die are dark. Reproduced with permission via Creative Commons (CC BY 3.0). Original color image was converted to black and white [2].

limited to an electrical and thermal test, where the wafer is held on a temperature-controlled chuck, and the devices are electrically contacted and measured. Sensors and actuators often interact with phenomena outside of electrical and thermal domains, such as motion, light, or acoustic energy.

For example, an accelerometer can be wafer probed for its electrical aspects such as opens, shorts, resistances, and capacitances. Some mechanical aspects may also be wafer probed, such as resonant frequencies, by electrically driving the device in mechanical resonance and measuring its frequency. However, wafer probe cannot measure the accelerometer's sensitivity. To measure sensitivity, the wafer prober would need to apply acceleration and measure the accelerometer's response. At the time of this writing, there is no known wafer prober that can do this. For other types of MEMS devices, there are a few specialty wafer probers than can apply pressure to pressure sensors by blowing a puff of air, or stimulate microphones with a speaker.

Despite wafer probe being incapable of fully evaluating MEMS sensors and actuators, wafer probe is still a valuable step in the overall manufacturing process. One of its main values is reducing the cost of back-end processing by removing electrically bad die before they go through the back end.

Package-Level Testing

The last typical manufacturing test for a MEMS component is package-level testing. It is done after your MEMS chip has been packaged, and it is the final test before the product is released to your customer. This is also the point at which the MEMS devices would be calibrated and compensated. Device calibration must be done after packaging because the package often influences the MEMS device performance.

Package testing checks the final MEMS product as a whole. Tests preceding package test may individually test the MEMS, ASIC, or package, but flaws could be introduced during the assembly that could cause the assembled product to be defective.

In package-level testing, a packaged device is electrically contacted and tested for functionality. For sensors, a physical stimulus is applied, and the sensor's response is measured. In actuators, an input signal is applied, and the actuator's output is measured.

When devices are tested for functionality, calibration and compensation are possible. Calibration and compensation are steps unique to sensors and actuators, which constitute the vast majority of MEMS devices. Calibration and compensation increase the accuracy and therefore the performance of the devices. It also imposes a uniformity of performance across devices.

Microfabricated devices can exhibit wide variation in their performance due to variations in manufacturing and assembly. For example, a critical dimension in an accelerometer is the gap between capacitive structures. The specification and tolerance for this gap may be 2 ± 0.5 μm. Half a micron is 25% of the 2 μm gap. Since accelerometer sensitivity scales as the gap dimension squared, the 25% variation in the gap results in a 56% variation in accelerometer sensitivity. In practice, with all sources of variation taken into account, it would not be unusual for accelerometer sensitivities to vary as much as 300% from device to device, which is untenable for a commercial product.

To reduce this variation, each device is calibrated and compensated. In order to perform a calibration, the device is compared to a calibration standard by stimulating the device with a known input and then measuring the output. In the case of an accelerometer, the device is typically measured in one orientation in the earth's gravitational field (+1 g), rotated to the opposite orientation (−1 g), and its response measured again.

Once the device's response is calibrated, it can then be compensated to impose a uniform response across all devices. The particular compensation constants for an individual device are recorded into the product's electronics. While the sensitivity of a population of uncompensated accelerometers may vary by as much as 300%, after compensation, the variation can be minimized to under 0.1% to meet customer requirements.

Trade-Offs in Manufacturing Test

Decisions about what to test must balance cost, time, completeness of data, consequences to the customer, liability exposure, and many other factors. Moreover, this balance must be continuously evaluated over time and the manufacturing tests adjusted. We discuss some of the trade-offs that are commonly encountered in manufacturing test.

Completeness of Data Versus Cost and Time

A significant trade-off to consider in manufacturing test is that the earlier the data is collected in the device's manufacturing process, the more cost-effective it is to dispose of a defective device. However, the test data gathered earlier is usually of lower quality due to the unfinished state of the device, and its accuracy in predicting the device's ability to pass package testing at the end of manufacture is lower (see Fig. 20.4). Conversely, the highest quality and most accurate information is gathered at the end of the manufacturing process, when the device has been completed, packaged, calibrated, and fully tested, but unfortunately this is the costliest point at which to remove a defective device.

There is also a general trade-off between completeness of data and the time required to gather data. The more thorough and complete the data collection, the more accurate the evaluation, but also the slower and more expensive the testing. Balancing this trade-off is several factors, including:

- The importance of the measured quantity.
- The cost of the measurement.
- The maturity of the manufacturing process.

For process monitoring, some parameters are more critical than others, for example, parameters associated with process steps that have a small process window. As a result, attention should be paid to these process parameters and detailed

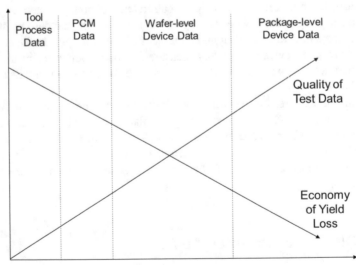

Fig. 20.4 Conceptual illustration of the trade-off between quality of test data vs. economy of yield loss. It is more economical to gather data and remove bad devices earlier in a manufacturing process; however, the data tend to be of lower quality due to the unfinished state of the devices

measurements taken frequently. For steps where the process window is wide, fewer measurements may be needed to have confidence in the yield of the overall process.

Some process parameters and measurements are easier to monitor and perform than others, and are therefore cheaper and faster to do. A parameter like electrical resistance is often fast and inexpensive to measure. But measurements that are more expensive and time-consuming to obtain should be used strategically and selectively. An example is measuring temperature hysteresis of a sensor. For temperature hysteresis, measurements must be taken at a minimum of three temperatures. It takes time for the device to reach thermal equilibrium, and test time adds test cost. Including a temperature hysteresis measurement as part of manufacturing test needs to be evaluated against the cost of the test, the frequency of the defect, the consequences to the customer, and many other factors.

Cost of Testing for Quality Control

More test accuracy takes more time, people, and money. As a result, manufacturing tests must be sufficient to ensure the quality of your product, but not so extensive that you would unnecessarily reduce your profit margin. Table 20.2 summarizes order of magnitude costs for different types of tests to help you consider some cost/benefit trade-offs. More accurate estimates of test costs can be obtained with identification of a particular test and its required resources. Test costs can add significantly to per-unit product cost and therefore must inform the product cost model (see Chap. 5).

It is ultimately a business decision to determine how much testing to do. Would a 1% defect rate be acceptable for your product? 0.1%? 0.00001%? The lower the defect rate desired, generally, the higher the cost of testing. Moreover, decreasing already low defect rates (e.g., <1 ppm) may have diminishing returns; it gets more difficult and more expensive to further reduce the defect rate because the failure mechanisms become rarer and more difficult to detect (see Chap. 14). If your product or market requires significant testing that your unit cost model or market price cannot sustain, then you will need to find an alternative solution such as altering the design, process, or software (discussed below).

Table 20.2 Typical test types, test locations in the manufacturing supply chain, and order of magnitude test costs in US dollars

Test	Location	Cost
Process monitoring	Foundry	Included in wafer cost
PCM test	Foundry	Included in wafer cost
PCM test	MEMS company or wafer probe service vendor	$100 s/wafer
Wafer probe	Assembly house or wafer probe service vendor	$100 s–$1000s/wafer
Package test, calibration, and compensation	MEMS company or test vendor	Machinery: $1Ms (NRE); $0.01 s–$10s per unit

Table 20.3 Decision matrix on when to implement manufacturing test

		Defect severity	
		Low severity	High severity
Defect frequency	Frequent	Case-by-case decision	Testing required
	Rare	No test	Case-by-case decision

Whether or not a test is essential to quality control depends on the consequences of the defect, the ease of detecting it, and the frequency of occurrence, as shown in Table 20.3. These factors, along with the cost, can help you to evaluate the cost vs. benefit trade-off. Consider these examples:

- A catastrophic, frequent defect that is cheap and simple to detect during test is a straightforward decision to make: implement the test and incur its cost.
- A rare, minor defect (such as an annoyance) that is difficult and expensive to detect is another straightforward decision: do not implement the test, and let any product issues be handled through customer service and warranty return.
- What about a rare, major (short of catastrophic) defect that is expensive to detect? Or a frequent but minor defect that is cheap and easy to detect? These should be decided case-by-case.

Case-by-case decisions should take into account factors such as the cost of the test, how the defect affects the customer, the end use market, market conditions, and business model. For example, some customers may have a system-level solution to deal with your product's defect(s), such as fault-tolerant software or built-in redundancies. Other customers using the MEMS product in a critical end use application may need the defect to be definitively solved by your company.

Managing Risks in Manufacturing Test

There are many technical issues in MEMS test that are beyond the scope of this book. For the purposes of this discussion, we discuss two broad, general risks that can affect manufacturing test at all levels and should be considered during MEMS product development.

First, beware of the mindset that there is a single, static, optimal set of tests for your MEMS product. The optimal set of tests will change as your understanding of the device matures, design and software is revised, the manufacturing process improves, customer and market needs change, the supply chain changes, test capabilities improve, and so on. As a result, the trade-off of test cost versus benefit should be continuously revisited. Your company should integrate this into its culture and understand that there is no best or final set of tests.

For example, as a manufacturing process matures through years of experience, the number of monitored parameters will change. We can illustrate this point by building upon the temperature hysteresis example from above. Despite the cost of the temperature hysteresis test, it was implemented because many devices failed this test. Over time, the manufacturing process improved, the root cause of the problem was solved, and eventually only very few devices failed temperature hysteresis testing. At this point, the necessity of the hysteresis test should be revisited and evaluated in light of several factors, such as the potential consequences for the customer of using a defective device.

Second, development and/or procurement of manufacturing test equipment is expensive and tends to have a long lead time, sometimes more than a year. If ordered too early in the MEMS product development, then there is a risk that the product test needs will have evolved and the equipment would not meet the new needs. One particular trade-off to keep in mind is that the higher throughput a testing system, the more customized it tends be, and therefore the less flexible it tends be.

It is, therefore, useful to approach high-throughput testing equipment in stages. For example, if the eventual business goal is to sell one million devices per year, interim stages might include an initial test system that handles 10,000 devices per year and a second iteration that handles 100,000 devices per year. As your understanding of the product and its test needs mature through these lower volume systems, the risk of mis-specifying a more expensive, faster, more specialized test systems is lowered. Once higher volume test systems are in place, the older test systems are still useful to the development team for failure analysis, debugging, and second-generation product development.

Strategies to Address Device Defects

Manufacturing test is only one method of eliminating defective products. For example, software may be a means to address or mitigate the occurrence of a MEMS defect, but it may not address the underlying source of a hardware problem. Additionally, updating the MEMS design and fabrication process has the potential to eliminate a defect entirely, but it is often the slowest and costliest to implement.

There are multiple solutions and strategies for managing device defects, and these can work together in various combinations. For example, when a new defect is identified, a software patch could be implemented as a quick, temporary solution to address the problem in the field. In the meantime, a new manufacturing test could be added at the factory in order to remove components with this new defect before shipment to the customer. Some failing components are sent to the failure analysis team to better isolate and understand the defect. Once failure analysis is completed and the defect is understood, a design revision or a process update can be implemented in the manufacturing line in order to eliminate the defect completely.

It is not always the case that the optimal solution to address a defect is a design update or process change. The most appropriate solution will depend on multiple factors: the nature of the defect, its consequences for the customer, the cost to implement a fix, as well as the company's business model, brand identity, and strategy. An acceptable solution may be applying a software patch and shipping the defective part anyway, or using manufacturing test to remove a defective part from the line and continue to incur the sunk production cost because it is cheaper than the design or process change. In some cases, when a defect is rare and/or inconsequential, it may be acceptable to not fix the problem at all and manage defective parts through a warranty return policy.

Summary

Manufacturing test monitors the quality of the entire MEMS product manufacturing process. Its implementation requires significant investment and planning, and its trade-offs must be thoughtfully balanced. It can be used as a powerful tool to improve product quality, increase manufacturing yields, lower costs, and to inform next-generation product design. Manufacturing test also impacts business strategy, such as product unit cost targets, product differentiation, and brand identity, and ultimately, customer satisfaction and business success.

References

1. Cassard J et al (2013) Standard Reference Materials® User's Guide for RM 8096 and 8097: The MEMS 5-in-1, 2013 Edition. National Institute of Standards and Technology Special Publication 260-177
2. Hsu C (2015) Clustering ensemble for identifying defective wafer bin map in semiconductor manufacturing. Math Probl Eng 2015:707358

Index

Printed in the United States
by Baker & Taylor Publisher Services